CITIES ARE ABNORMAL

CITIES ARE

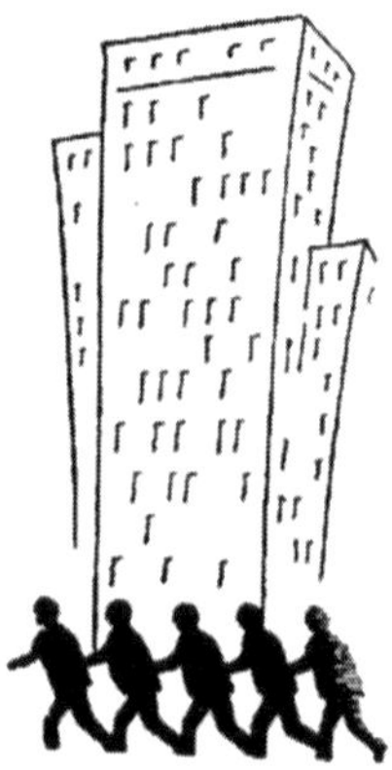

ABNORMAL

Edited by ELMER T. PETERSON

Louis Bromfield

Jonathan Forman

Ladd Haystead

Henry L. Kamphoefner

S. C. McConahey

H. C. Nixon

J. J. Rhyne

Paul B. Sears

Roy L. Smith

Warren S. Thompson

Paul L. Vogt

ISBN: 978-1-940001-65-4

Originally published in 1946

Sportsman's Vintage Press
2017

Foreword

EMPIRES and commonwealths are born of farms. Must they die of cities? The amazing year 1875 was at once a pivot and an usher of great events for the United States. One event was the Philadelphia Centennial Exposition, a milestone in our industrial progress. Telephone, electric trolley cars and lights, the agricultural settlement of the Great West, the exploitation of forest and mine–such were the more notable achievements which came in quick succession, to be followed by countless other inventions and enterprises which we today have at our disposal.

These were things for all to see. But within the impersonal array of census figures was the evidence of something still more fateful–a revolution behind a revolution. During the decade of the eighteen seventies the population of this great agricultural nation became predominantly urban. Nourishment continued to come from the farms but the growth which it fed flowered in cities. The expressions of urban activity became overwhelmingly vocal. Not even Bryan's silver tongue could prevail against them. The castigation of the Seven Devils of Wall Street by the Populists Peffer and Weaver left no mark upon a brashly growing urban pattern. The forms and folklore of city living began to set the mold of our thoughts and aspirations. Rural education, and even higher education in rural states, became a sort of pedagogical cream-separator, pouring the richest of its products into the stream of urban life.

Superficially this might seem to be nothing more than what had happened often enough before, although on a far smaller scale. It had happened in Rome between the days of

Cato and those of Augustus, in Palestine between the time of David and that of the prophets. We still do not know whether this change, in these places and elsewhere, was the symptom or the cause of decay. But we do know that never before our own day had the process of urbanization been facilitated by such instruments as we have. No man can yet say whether we possess them or are possessed by them.

The city today is not merely an aggregation of men and women with highly specialized functions. In addition, by virtue of modern technology, it represents an engulfing process of standardizing multiplication. It does not confine itself within bounds. The great forcing pens of genetically uniform and scientifically fed domestic animals, vast plantings of wheat, cotton, and fruit, whose products are handled on an assembly-line basis—what are these but a penetration of the urban pattern into rural life?

We are forced to take notice of many consequences of this concentration and forcing; for instance, we know that they greatly increase the dangers of animal and plant disease by reason of crowding and unnatural environment.

Nature gives ample notice of her abhorrence of such interference with her own system of what superficially seems to be heterogeneous placement, but which actually is orderly, because it distributes risk in such matters as reproduction and health, instead of being governed by rectangles, abstracts of title, or geographical and political patterns. She refuses to put all her eggs in one basket. She operates no central incubator or commissary.

Furthermore, nature has never given notice that she has set mankind apart, to be governed by a set of rules wholly different from those governing all other living things.

Within the city, intensity of conflict, the dangers of pestilence and of economic disaster have been multiplied. Human mutual relationships and valid, well-rounded personal development become more difficult. Within the urbanized, mass-production rural areas new problems of unbalance, of epidemic, and of the destructive enemies of animal life have arisen. Our scientific advances in agriculture have not been able to offset these consequences. In at least one great agricultural state, such advances have merely served to maintain, without increase, the average per-acre yield.

We have employed science with an amazing cleverness and facility to accomplish the things we wanted most immediately, but without sufficient thought of the larger course of scientific compulsion within which we must work. In this respect we have not been unlike a man who scrupulously observes municipal ordinances while doing violence to national statutes and the great tradition of the common law.

Scientists are fairly well agreed on the order of magnitude of the greater landmarks in earth history. Our records of life go back certainly well over a billion years. The age of humanity is not more than a few thousandths of that interval, by the longest reckoning. Conceding man to be at least a million years old, the period of modern urbanization in our country, from 1875 to the present, represents the kind of infinitesimal fraction that is safely ignored in the beautiful operations of the calculus.

True enough, the life of man is measured by the quality of his experience and not by the clock. But in nature the intervals of time have a meaning which we cannot neglect. During the enormously long time before man's coming, through trial and error and survival, the relationships of living beings to the world about them were established. We have to take these relationships as we find them and build upon them. They apply to ourselves no less than to the plants and animals which serve our purposes. We must look to these principles for our first predication in all human enterprise. Yet there are those who would cast all education, all living, and all values into the mold of modern urban experience.

It is convenient for historians to judge vast human movements by sociological, economic, or political criteria. It seems to be a bumptious human characteristic to assume that a human world is literally and in all respects self-governing. In one sense this may be true, but only through the agencies of stern biological processes which humanity is prone to disregard. A concrete illustration of this misconception of history is furnished by Dr. W. C. Lowdermilk, who asserts that it was the loss of fertile topsoil through human ineptitude which caused the downfall of the ancient and once powerful empire of Babylon; yet the convenient historical explanation has to do with battles, dynasties, political changes, and other purely human conflicts or relationships.

Ultimate order for mankind is not simply a matter of true rectangles and flawless bearing surfaces, important as these may be. It is rather a problem of respecting the balance of nature. If violated, this balance will right itself and prevail, however drastic the consequences to any organism. Among animals, contagious abortion, for instance, would soon disappear in a natural, heterogeneous environment, because it is self-isolating and self-extinguishing, whereas, in an urbanized environment, it is invited to wreak its havoc in a widening circle. Among humans, the congestion-born pestilence is a familiar phenomenon.

The authors and editors of this book hope that the reader will view what they have to say against its deepest visible background. This background is the long and deliberate course of nature's process rather than the brief course of our national history, or even the tiny splinter of time represented in the history of human civilization. It is their hope that in this great perspective the reaction of human events to order will appear. This, after all, is a more hopeful quest than the search for rigid, square-by-square order in history.

The increasing complexity of our economic-social system, thrown into still greater confusion by two global wars, with all their terrific wrenching of human passions, with consequent swaying of judgment, has led to a phenomenon never seen before, except possibly in the regime of the Sophists, who preceded and probably furthered the debacle of the glorious Greek democracy. These Sophists suffered acutely from ingrowing brains and Alcibiadean versions of humor.

There is no dearth of information. In fact the wealth of information has become so embarrassing in its bewildering surfeit, and it is so voluminous, that no one man or group can hope to assimilate a well-balanced picture of our situation, when a nation as large as ours is considered.

The net result of this perplexing phenomenon is that important individuals and groups, particularly in the densely populated regions where activity is most frantic and intellectual impacts are most bewildering in number and variety, have accumulated a broad and accurate knowledge of things that aren't so.

It seems that confusion increases in geometrical ratio with the opportunity for learning. Larger and larger areas become afflicted by a folklore which deliberately and unrealistically isolates itself from the natural world.

One manifestation of this is the growing dependence of the average man upon the collective resource—if, indeed, it can be called resource. It takes the form of a heart-felt hope that government can by some legerdemain abolish insecurity and by the very nature of collective action achieve a balance in human affairs. Such thought leaves wholly out of account the nature of the disease which the easy formula of government action is supposed to cure.

This is, of course, a characteristically urban concept, based upon a great fear, superinduced by the length of the supply line. It is based in part also upon the unanalyzed theory that everybody must work for somebody else, and hence be dependent upon somebody else.

One of the great hopes indulged by those of us who have

contributed to this book is that it will offer some measure of diagnosis for the ills which beset us domestically. If we shall have accomplished this purpose, then we have good reason for believing that decentralization will in time be realized in the United States, with results that cannot but be beneficial to an economic and social structure that has hitherto been characterized by chaos and immaturity.

In editing the chapters written by men who have developed their own independent versions of the problem in hand, I have neither suggested what they should write nor tried to make them conform to a preconceived pattern. It seems to me that the spontaneity of the various presentations is far more significant than conformity could be. To preserve this spontaneity I have therefore winked at a certain very small amount of duplication, because eliminating it might detract from the unity and coherence of the individual theses.

The very slight amount of duplication might, in fact, be considered an element of strength, in that it shows that when multi-angular approaches are made—those of the sociologist, biologist, ecologist, business economist, manufacturer's commentator, medical authority, spiritual leader, humanitarian, and so on—there is unique potency in the fact that two or perhaps more men of widely varying interests and occupations have arrived at identical though independent conclusions. I feel sure, therefore, that these rare and minor overlappings will not be taken amiss. The fable of the six wise men of Hindustan, each of whom gave a different version of what an elephant is like, does not altogether apply to this book.

There are a few minor aspects wherein conflicting opinions, or opinions not shared by myself as editor, are expressed. Again it is felt that spontaneity is more valuable than conformity, and the remarkable thing is that there is indeed no lack of agreement on any material aspect of the main problem.

I feel exceedingly proud of the eminent array of writers represented in this book, enlisted in co-operation with my publishers, the University of Oklahoma Press. Each one of them is a recognized authority in his field. Indeed, I feel presumptuous in putting my name in the same hat with theirs. My only excuse is that I have tried, as a working journalist, to tie the various subjects together by means of a layman's point of view. In all humility, I hereby express my heartfelt thanks for their distinguished contributions.

ELMER T. PETERSON

Contents

Notes on the Contributors

Elmer T. Peterson, the editor of this volume, is the author of *Forward to the Land* (1942). Formerly editor of *Better Homes and Gardens,* he is now an associate editor of the *Daily Oklahoman* and the *Oklahoma City Times.* Articles by him on a wide variety of subjects have appeared in magazines and journals published in this country.

Louis Bromfield is one of the best known of American novelists and playwrights, a Pulitzer prize winner in letters, and the successful owner-manager of Malabar Farm, Lucas, Ohio. His most recent book is *Pleasant Valley* (1945).

Dr. Jonathan Forman, a practicing physician in Columbus, Ohio, specializes in allergy and is interested in related subjects, such as community health and soil conservation. He has for a number of years been editor of *The Ohio State Medical Journal* and a member of the editorial board of *The Land.*

Ladd Haystead, farm editor of *Fortune,* has written widely on agricultural and rural subjects. He is the author of *Farm for Fortune and Vice Versa* (1942), *Preacher's Kid* (1942), *Meet the Farmers* (1944), and *If the Prospect Pleases: The West the Guidebooks Never Mention* (1945).

Henry L. Kamphoefner is professor of architecture in the University of Oklahoma. His articles have appeared in a large number of technical and professional journals since 1934.

S. C. McConahey was, until his retirement at the end of 1945, the vice president and treasurer of the Westinghouse Air Brake Company. He has devoted a considerable period of study to the problems of urban-rural relationships.

H. C. Nixon is a lecturer in social science at Vanderbilt University. He was a contributor to *I'll Take My Stand* (1930) and *Culture in the South* (1934), and is the author of *Forty Acres and Steel Mules* (1938) and *Possum Trot* (1941).

J. J. Rhyne has been director of the School of Social Work

in the University of Oklahoma since 1929. He is the author of *Some Southern Cotton Mill Workers and Their Villages* (1930) and of articles on social and cultural problems.

Paul B. Sears is one of America's best-known botanists, the author of *Deserts on the March* (1935), *This Is Our World* (1937), *Who Are These Americans* (1939), *Life and Environment* (1939), and *This Useful World* (1941). He is professor of botany in Oberlin College.

Roy L. Smith, D.D., is editor of the *Christian Advocate*, the official magazine of the Methodist Church, and has for many years been a leading American churchman.

Warren S. Thompson has been since 1922 the director of the Scripps Foundation for Research in Population Problems, Miami University, Oxford, Ohio. He is the author or co-author of seven books and of numerous articles and pamphlets on population subjects.

Paul L. Vogt is senior social scientist, United States Department of Agriculture. His book publications range from industrial subjects to sociology and rural economics, and his articles have appeared in the principal magazines and journals devoted to economics, sociology, and city planning.

CITIES ARE ABNORMAL

1. *Cities Are Abnormal*

ELMER T. PETERSON

THE American metropolis, to which an uninterrupted increase in population has seemed the very breath and law of life by reason of the traditions of boosterism, now stands hesitant and uncertain.

In the fifth decade of the twentieth century, the city finds itself no longer the master of its own destiny, because it has set in motion certain forces and trends that it is unwilling or unable to stop. Forty years of almost precipitate growth, accelerated by two wars, have brought the average metropolitan center to the stage of bursting its seams. The best laid plans of chambers of commerce and planning commissions have proved unavailing against an erratic dynamism that has shuffled populations, redistributed them to meet the partially war-activated demands of an enormously expanded industry, and set at naught the orderly processes of development so dear to the hearts of Americans.

There is a vague uneasiness. By some it is asked, "Is this thing real?" By others, "Can the pace be maintained if and when the world settles down to lasting peace?" and "How about the population balance between city and country?" The white-collar worker has wondered if it would not be wise to acquire five or ten acres at the edge of the city, against a rainy day, and the banker has long since discovered the trend of things. Many of his clients are buying in the country—"It looks like a secure thing," they tell him. "My children will grow up normally there. I can get the kind of recreation I haven't had in a dozen years, and besides I can live comfortably there when I retire."

The wisdom of the people is intuitive but seldom articulate, and it often awaits scientific corroboration. The uneasiness of millions of Americans about metropolitan life is the result of unguided but sound questioning of things which have heretofore implied material security—concentration of large-scale industry, the apparently stable flow of commerce to river and port towns, the increasing power of those cities which, favored by history or circumstance, or by statute or trade practice, have acquired financial domination of vast producing areas.

The uneasiness is not a thing that was developed with the quick growth of war industry alone. There are half-remembered incidents from the period of demobilization of men and industry after World War I. Of bread lines running three times around Times Square in 1931. Of unstable conditions even in the midst of plenty during World War II, when New York City itself at one time had an unemployed population of nearly half a million, because industry had favored other areas over the financial and commercial giant of the Western world.

Even greater and more persuasive than material security, however, is the desire for spiritual peace. Nearly two thousand years ago Horace wrote the fable of the town mouse and the country mouse, perpetuated in nursery tales. He was merely repeating an instinctive expression that was heard even in primitive times, when a city was chiefly an armed and walled camp where people huddled together for mutual protection. "Better a thrifty condition and happiness than wealth and discontent." Most curiously, the average American, with the homeliest and sanest of desires, is least able to satisfy them. Room to live in, fresh air, flowers, creatures for the children to know and enjoy, simple, sane recreational and social opportunities for the whole family—ideals which have survived from the dimly remembered days of the small town. But the attainment of these without the loss of those things which a mighty industrial civilization has produced—the theater, music, the arts, alert and progressive schools, hospitals, sanitation, scientific research—this is the dilemma of twentieth-century America.

One of the axioms of political science is that no movement is ever unmixed. Its premise is that the force of any movement always carries beyond the conditions which brought it into being. The American city began largely because of two requirements of industrial production and commercial activity: motivating power in small area and short-line transportation for workers. But the pattern laid down in a relatively primitive age of power and transportation has survived to the twentieth century, an age which might logically have been expected either to have improved upon the pattern or to have dissociated itself from it entirely.

The long arm of electricity—a modern and therefore not fully understood factor—has wrenched our industrial economy into a new and highly elastic form. Here we need not talk in abstractions, for the physical aspect is as plain as an egg beater. Electricity has performed this feat because it can operate anything from a huge stone crusher to a barber's clipper at a distance of anywhere from ten feet to three hundred miles, whereas the steam-driven power shaft which started the Industrial Revolution had an efficient radius of only a few score feet.

Electricity has other implications that may prove even vaster and more pervasive in reinforcing this trend—the motivation of instantaneous communications, television for example. The general accomplishment of the electrical djinn in this relation is that he briskly changes the congested, perpendicular concept of a city into a challenging potential of far-flung, sprawling, horizontal shapes which disdain geographical limitations.

This furnishes one excellent reason for refusing to assume, merely because the Industrial Revolution set in motion certain trends, that those original trends will continue their momentum. True modernism—the progressive outlook—does not consist in the ability or tendency to continue a given force in a straight line, but in the genius for adapting oneself to unpredictable change.

To illustrate this point, a high-school class in a large middle western city was asked to put its imaginative young artists to work picturing the city of fifty years hence. The youngsters were well informed and alert. They had acquired a psychological background of contemporary science, sociology, and economics, and in that sense were modernists, but they had not sufficiently heeded the voice of experience, which would have told them that the chances are against a straight-line movement of present trends.

They pictured larger and larger skyscrapers, greater and greater speeds in transportation, greater urbanization—in short, merely an intensification of what we now have.

Working along similar lines, a metropolitan architect-artist pictured the ideal summer resort of the year 1980 as a monster fifty-story hotel, on the roof of which is a huge landing field for airplanes, with various floors devoted to ballrooms, cocktail lounges, and so on. All of this on an ocean beach. This, of course, is merely a trite repetition and intensification of the present pattern, with no essentially new motif. It did not seem to occur to him that the ideal summer resort of 1980 might be situated on bypaths of the present onrushing juggernaut of urban-inspired happiness-seeking. It might conceivably be a decentralized though interrelated series of seaside or mountain or lakeside units, each of which might be instantaneously and completely served by the sights and sounds of every important contemporary event, so situated that the occupants might reach a central amusement factor within a few minutes. Thereby might be preserved the actual or putative advantages of both urban and natural ecologies.

The vertical movement has carried far beyond the requirements of physical power created by turbogenerators, diesel engines, and other modern energy producers. It has gone on and on by inertia and habit-thinking. The forecaster of the future will be deceived if he assumes that this movement will continue indefinitely. There are valid evidences that the Empire State Building is both literally and figuratively the crest of the vertical wave. Significantly this building is a monument to congestion.

In the language of physics, the law of inertia applies as well to bodies in motion as to bodies that are stationary. Sometimes a great war or drastic social upheaval is required to remove the peculiar inertia that confines itself to present patterns, a fantastic but logical extension of which is a five-hundred-story building. If we are to give respectful attention to incisive and fearless commentators like Mumford, Spengler, Geddes, and Ferrero, we are likely to conclude that, because of a cyclical pattern, the inertia of movement may prove even more undesirable in the long view than the inertia of relative immobility, such as has been seen, for example, in the once static Chinese civilization. That is, unless we adhere to the philosophy of the ephemera—the fly that lives but a day—"It was a great life while it lasted!"

Urbanism seems necessarily integrated with the ideal of money-wealth. The dollar economy is supreme, for the easily understandable reason that urbanism is the product of intensive specialization, and there is no method whereby specialization can flourish except by the techniques of money. Ruralism and, to some extent, decentralization point unerringly toward the wealth of goods and services—the wealth of production—the economy of abundance. This contrast deserves special attention, to be given later.

A high executive of a great life insurance company has remarked that in his opinion the mere money-getting genius of big-city life is a false basis upon which to create a civilization; that it is thoroughly conceivable that a good public-school teacher might justly be rewarded more highly than a financial manipulator. We may be a long way from the realization of this ideal, but it is worth thinking about.

Stating the proposition in a different way, the vice president of one of the largest industrial concerns in the United States declares quite emphatically that the urban-rural ratio of population, which was about 75 to 25 before World War II, is a high explosive which threatens to destroy our economy. He says that this unbalance is the basic cause of all of our major economic problems. The war, by drawing off farm manpower into the armed services and war industry, changed this ratio to the still more dangerous figure of 85 to 15. This lopsided ratio puts the burden more and more upon a dollar economy, a shaky dependence. Translated into practical terms, it means considering the farm as a service station, operated by urban-based day laborers who work for dollars. In such a system, eighty-five urbanites figuratively stand in line on pavements, waiting for the fifteen rural men to feed them, and sometimes assume, moreover, that it is the duty of the taxpayers to keep them on pavements and away from the disagreeable task of raising their own food.

Food rationing opened the eyes of many. People may realize that no matter how disagreeable it may be to hoe potatoes or milk cows or slop hogs, we are only a short distance ahead of starvation, even in this great productive country, and food is by far the most important material thing we may have, and primary production from the soil is our best index of wealth. Under rationing the $100,000-a-year man gets no more than the $1,000-a-year man. In fact the latter gets even more than his wealthy neighbor if he operates his own food garden or farm. If you want to see evidence of real wealth, attend a farm picnic in some remote rural district and consider the mounds of fried chicken, ham, cake, and pies produced without benefit of money.

As a result of this change of thinking, an important change in conclusions is under way as this is being written. About two years previously the principal editor of one of the most important metropolitan magazines wrote a letter to the writer in which he defended the idea of commercial farming and with it the dollar economy of the National Resources Board public works program, expressed in "jobs" on day wages. He went so far as to say that to put a city worker on his own food farm would be a bad thing because it would "take him out of the economy."

Fortunately this astonishing point of view was not long maintained. Several months later, he wrote an able, eloquent, and convincing appraisal of the American economic future in which he apparently changed his views completely. He boldly declared that the best friend and exemplar of economic freedom in America is the small, new, ambitious, competitive-minded enterpriser—the man who wants to be in business for himself. He added that the farmer is the freest of men, because he produces his own food, and "wealth is not the end of life," meaning, of course, money. He concluded by asserting that a true policy of freedom is one which first sets men free to make a living.

If it seems incongruous to mingle references to food farming with statements about free, small industrial enterprises, it should be remembered that the farmer is the original free-enterprise industrialist, dealing with the most substantial form of wealth.

During the war period, the alternatives of collectivism and finance capitalism loomed before the nation. The editor to whom reference is made does not seem to think that these are the only alternatives—in fact he envisages a rebirth of the Adam Smith formula by way of small, decentralized projects.

The metropolis, whether it operates through collectivism or finance capitalism, is essentially a huge mechanism for interdependence, and it operates against the philosophy of Adam Smith, which was contemporary with our own Declaration of Independence and Thomas Jefferson's prophetic, and now, it seems, ultra-modern philosophy.

Interdependence, like other aspects of the Industrial Revolution, has encountered the law of diminishing returns. Being pushed around in a great metropolis is getting to be quite annoying to many people. They are looking about for remedies. As our editor pertinently says, "economic freedom is becoming scarcer." It becomes scarcer as congestion and economic centralization become more abundant.

An infant rebellion is born in the mind of every thinking American when he is jammed into an elevator car or is herded into a subway train, or is forced to stand in line to be served. It doesn't seem right, but it is the price he pays for the doubtful blessings of what he calls civilization, which, he is told, is getting to be frightfully complex. He doesn't know exactly what to do about it. That is why he is in a state of uncertain rebellion, which, happily, does not take the direction of many rebellions—toward collectivism, but directly the opposite—toward a greater individualism.

The metropolis need not be a prey to racketeering and political corruption, but in practice it has often been so. As a matter of fact, the condition of our cities in these respects has been cause for continuous alarm at home and the occasion for startling and often uncomplimentary observations abroad. The simple truth is that, given a congested urban area and a passive democratic electorate, crime and peculation can be made to pay. The city has witnessed the passing of two of the most important sources of socially healthy action in the disappearance of the neighborhood and the community. Without them, democracy operates haltingly and often by remote control, which really means no democracy at all.

Congestion, when viewed on other than moral, esthetic, or economic grounds, seems to be based upon a vague and primitive instinct for huddling in tense groups. This huddling, under modern conditions, is seldom justified—in fact it would probably seem inexplicable to a man from some other solar system whose ancestors had not passed through the habituated folklore of a hundred centuries of primitive warfare and other dangers and deprivations. It was originally an emergency technique rather than an expression of normal social craving. We may rightly ask whether we are unnecessarily prolonging an emergency—does not modern warfare (specifically the atomic bomb) advise us to scatter rather than to gather in dense clusters of humanity?

From almost every angle that we view urban life in America, the decentralization of cities seems desirable—public health, economic betterment, economic logistics, moral welfare, better local utilization of natural resources, better distribution of manufactured products, a better conceived military defense, a more rational architecture, and, in general, a happier adaptation to the changing mores.

In opposition to decentralization, it has been argued that our very civilization has depended upon the leadership and stimulus of cities, and that the propinquity of many specialists has proved advantageous, both for the perfection of technology and for service to the public. It may also be argued that cultural, religious, and entertainment organisms have flourished best where large populations have congregated in compact groups, that short distances are advantageous in transportation and other vital economic facilities, and, in general, that the gregarious instinct of mankind is exercised logically in the complex mechanisms of urban life, involving social welfare, education, frequent stimulating contacts, and convenience in all manner of transactions.

In view of such conflicting arguments, it is desirable to make analyses as calmly and scientifically as may be possible, to raise various questions and explore their answers, and to consider implications that may readily occur to the reader.

Just what do we mean when we speak of a city? Do we mean something that is measured in terms of so many square miles of land? Is it chiefly a physical and geographical concept, for example, a group of skyscrapers covering ten blocks and a belt of shops covering a dozen more blocks, a layer of apartment and boarding houses, spots of suburban trading centers or theatrical projects or industrial plants, and a wide encirclement of individual residences? Just what is there, besides habit-thinking and folklore, that ordains this pattern?

On the other hand, are we considering the city primarily as a community of human beings trying to live the best possible life?

Your answer to this choice of meanings is exceedingly important.

To put the question in another way, do we consider the city in the light of individual aspirations—the preservation, in so far as possible, of personal dignity, liberty, privacy, and the gratification of the individual's hunger for companionship or entertainment or religion or the touch of earth and sky, or whatever else he may chiefly crave? And when we stress conveniences, do we really produce convenience, or its opposite?

Of course it is necessary first to determine, if we can, what we think we are living for—what values we deem important. To do this, we may find it sometimes necessary to weigh durable satisfactions against various sensory thrills and stimuli which prove to be short-lived, perhaps not for the purpose of excluding one or the other, but with our eyes open—with scientific knowledge of what each goal of desire actually means.

One of the primary corollaries that seem to develop from this premise is that the ideal community, or design for living, or way of life, or whatever you may wish to call it, is something that is extremely difficult if not impossible to blueprint. It seems automatically to fall into an arrangement in which no two individuals view their best satisfactions in identical terms.

A basically standardized Utopia, it seems, is outside our purview. We have abundant warnings, in the form of experiments like the Oneida Community of New York state, Amana, Iowa, or New Harmony, Indiana, that personal desires are so persistent and unpredictable that they resist blueprinting. If there is to be a vision of an ideal decentralized city, it will have to be created in the mind of each individual separately, though subject to a few simple and obvious community requirements.

On the other hand, of course, city, district, and regional planning and efficiency correlations are essential to healthy decentralization. The whole subject is pragmatic and adaptive. It need not be assumed that the convinced decentralist must advocate a chaos of incoherent and unrelated individual whims and caprices.

To put the balanced proposition in practical everyday terms, it could be stated that nobody knows precisely how to plan a vast decentralized city because nobody knows how many Baptist churches or saloons or bowling alleys or filling stations or colleges to specify, or where they should be placed. Such elements depend entirely upon human and ecological variables.

At this point it is desirable to say that for the sake of convenience, and to save time, the word "urban" in this book is intended to mean "congested urban," especially since experience has given it that connotation. As we proceed, nevertheless, we may project a new type of urbanism to form a contrast with the old.

Is urban life the norm of existence? Or, to coin a new word, is it an abnorm?

Going back to the question of whether we are to consider the city as a mechanical and geographical thing or as a group of human beings trying to find the best possible life, we encounter the fascinating subject of human ecology—the problem of environment.

There seems to be a growing folklore which presents a picture of urban life as being the "better life" which quasi-idealistic or politically-minded persons visualize as the rightful due of every person, and some even say that it is the duty of government, including rural taxpayers, to see that every person is guaranteed the supposed advantages of urban life, and, if necessary, is granted a state subvention, even if he is idle, so that he may remain in the city.

Along with this folklore travels the theory that the farm is merely a service station for urban living, and that farm life should be considered abnormal and unwholesome drudgery, to be avoided or got rid of as quickly as possible.

On the other hand there are those who argue that farming, particularly for food production, is the fundamental occupation, since it provides the shortest and most direct line of supply for living from the soil—which all of us do, even if we live in a city apartment. After all, of course, everybody lives from the soil, and the man who procures his food most directly from it comes the closest to being a sound and indestructible human economic unit. This is merely another way of saying that farming, as a way of life, is the safest and surest way of living, even though some of us may consider it laborious and disagreeable, and the advocates of this way of life use not only the economic argument, but the argument of health and general well-being. This point of view insists that rural ecology, despite all the tremendous advances and changes brought about by urban civilization, is still securely and permanently supreme, and they cite new scientific revelations to support their position.

The newness of this approach is high-lighted by the fact that there was no book on human ecology until the middle thirties. Since then, however, a great multitude of scientific facts have crowded into the picture, and they have been sharply pointed up by invention of devices that make rural ecology seem not only more practicable, but more conspicuously or dramatically normal.

Notwithstanding the fact that science has lagged in establishing the norm of human existence, imaginative philosophers have long cherished intuitions on this point. Nearly a century ago, Taine, in his history of English literature, said that the body of man in every country is deeply rooted in the soil of nature. About seventy years ago, in the *Kansas Magazine*, John J. Ingalls, statesman, essayist, and poet, wrote:

> He [Taine] might properly have declared that men were wholly rooted in the soil, and the character of nations, like that of forests, tubers and grains, is entirely determined by the climate and soil in which they germinate. Dogmas grow like potatoes. Creeds and carrots, catechisms and cabbages, tenets and turnips, religion and rutabagas, governments and grasses, all depend upon the dewpoint and the thermal range. Give the philosopher a handful of soil, the mean temperature and the rainfall, and his analysis would enable him to predict with absolute certainty the characteristics of the nation.

Must the true and irreducible values of the city be squeezed and forced up through conventional matrices into perpendicular phenomena? Is congestion an asset or a liability? Does physical propinquity of human beings necessarily result in spiritual sympathy or like-mindedness or good will?

Reverting to the design of living exemplified in individual food-farming units, may we visualize a rural city—a city that combines all the good features of both urban and rural environments? If we grant the arguments in favor of rural ecology, do these same arguments apply proportionately to the rural factory, the suburban office building, the village television theater or analogous decentralization features?

At this point someone is sure to interject the homely old question—can we eat our cake and have it? Perhaps we can. Let us explore further.

Approaching the problem from the urban point of view, we see enormous and unquestioned values in the Industrial Revolution, which began to show its power approximately as of 1790. This phenomenon became the basis of the modern city, whose chief elements are intensive specialization and an industrial, commercial, and financial underpinning, with the inevitable cultural, social, scientific, recreational, and other concomitants.

Before the Industrial Revolution, the city was a loose aggregation of individual and largely independent shopkeepers, artisans, artists, doctors, teachers, preachers, laborers, and so on. There were no corporations as we know them. The metropolis as we know it was nonexistent. After the advent of the Industrial Revolution, the city was changed into a tightly-knit and highly interdependent community of specialists, organized largely into compact groups of like economic or cultural aspirations and stratifications. In this sense the modern city is only a century old—a mere watch in the night as related to the long journey of mankind.

The Industrial Revolution, with its inevitable expression of modern urbanism, seems to have worked well up until the twentieth century, but now there are ominous signs that the law of diminishing returns is at work on it. We can no longer assume that certain motivations, trends, and patterns, because they worked well up to the year 1900, will continue, in straight-line momentum, to work well in the year 2000, with continued accretions of power, bulk, perpendicularity, congestion, speed, and other man-made attributes.

Man, after all, is the only creature who has changed his own ecology, and, incidentally, that of other organisms. His power to do this, however, is disconcertingly limited, as is almost everything else, by the law of diminishing returns. It is most curious that man has understood and been guided by this law at almost every point, save where the economic merges with the social. And man's most stupendous social creation is the modern metropolis, where the law of diminishing returns is operative, often little noticed, though with fateful consequences to man and his society. But the areas of diminishing returns are many and stubborn. Let us consider transportation, for example.

It has been said that the freight rate schedule began as a simple device for expediting manufacture and evolved into a monster which no one could manage and only the experts could understand. The present rates are a system in which appears the "long- and short-haul" evil—the system whereby a metropolitan center obtains a preferential transportation rate over the smaller manufacturing point along the same route with shorter haulage. That the point of diminishing returns, in this respect, has been reached long ago is patent. But the social view which recognizes this fact has, until quite recently, had little effect upon industrial practice. Gradually, however, it is becoming clear to industrialists themselves that society's loss under such a system is also industry's loss, and eventually the big city's loss, and that remedies must be found, and speedily. It doesn't pay, in the long run, to be a blood-sucker.

Thus, if the best and socially most economical use is to be made of our resources, many industries must be established, or even possibly re-established, near the point of origin of raw materials and the area of consumption. What this may mean in terms of reduced transportation is even now fairly calculable. A redistribution of industry is imperative to decentralization, not only of cities but of American economic life.

At one time water transportation was the most important element in the building of cities, and the important centers were invariably established on good ocean harbors or on navigable rivers. Because of inertia in the rut-line movement of population and commercial trends, this original factor still plays an important part, even though most of the traffic utilizes railroads, highway trucks, and airplanes. Here is another task for the economic realist who wishes to overcome habit-thinking and traditional patterns that persist despite logic and common sense.

The emergence of new plastics, light metals, chemurgic farming, and other modern economic factors should provide impetus and encouragement for such a realist, and postwar economic readjustments, involving the reorientation of a vast number of cornfield airplane plants, munitions works, and the like, should provide the opportunity.

While thinking of postwar implications, the prudent observer will not neglect the exigencies of war itself. It may be that war will come again. Each recurrence brings new inventions, some of them so diabolical in cleverness that the task of defense becomes more and more baffling. This provides all the more reason for the protection of defense industries and vital military nerve centers from attack. This big city is a wide-open target. Even with most elaborate camouflage, Berlin was unable to deceive the air raiders, while in the diversified landscape of the Ruhr, the attackers faced a stupendous task. In the future all vital installations will have to be concealed in rural areas, under natural camouflage.

Britain brought military decentralization to a high state of perfection, and this country must take notice of the lessons taught with fire and atomic bomb.

In appraising the trends toward decentralization there are imponderable as well as ponderable factors. The inspiration of the open spaces and the sight of mountains, forests, lakes, rivers, and sea, the opportunity to behold sunsets undefiled by smoke and sordid fabrications, the health-giving contact with the soil and grass and trees and flowers, companionship with wildlife—these are but suggestive advantages.

We who are called "civilized" may ridicule the unsophisticated savage, who bows down in reverence before the awful might of lightning, hurricane, thunder, or ocean wave; but if we ourselves bow down before skyscrapers or locomotives or gigantic steamships, we are less to be admired, for the savage is the one who pays homage to the powers that are infinitely greater than anything contrived by puny mankind. In fact there is a definite form of worship among many whose thinking is overpowered by the alleged might of the metropolis. The imponderables were well sensed by Alexander Pope, who wrote:

> *Lo, the poor Indian! whose untutored mind*
> *Sees God in clouds, or hears Him in the wind;*
> *His soul proud Science never taught to stray*
> *Far as the solar walk or milky way.*

We are all brought around sharply to face elemental facts from time to time, and while contemplating a war crisis, when all human foundations seem to totter, it is salutary to reorientate ourselves and make new appraisal of the general problem of living. At such times we seem to realize that there is wisdom among savages that we do not have ourselves—we who have worshiped cities with "ecstasies of power"—man-made power.

In considering the city as an abnormal entity, of course

we need not assert that all abnormality is undesirable, for there are certain abnormalities, like highly bred dairy cattle and enormous roses, that are the more valuable because they are abnormal. But we must always stand ready to pay the price for abnormality.

If man is able to survive the penalties of deviating from nature's ways, that is his business. But let him be warned that they are constantly hanging over him, and that the utmost vigilance and intelligence are needed in paying for immunity, since he may incur penalties even after he thinks he has completely circumvented kind but crafty nature.

The point of departure, in considering the growth and evolution of cities, is that natural or rural life is the inescapable norm. This thesis was once considered a sentimental or moral one. New facts provide new bases for appraisal. This book seeks to present facts, together with conclusions of eminent authorities in their respective fields, which argue for the decentralization of cities. Underlying all these indices—sociological, economic, logistic, architectural, moral, or psychological—is the one basic fact of ecology—the science of the relation of any given organism to its environment—for, as a great ecologist aptly says, "Life and environment are thoroughly interwoven."

We may have to wrench our attitudes if we would prepare ourselves for a coolly scientific approach. For thousands of years the human race has assumed, with an air of smug finality, that its Utopia is to be achieved by men's learning to live with one another. In other words, the ideal social system, no matter what form it may take, has been considered the criterion and predicate of a perfect design for living, so much so, in fact, that the two values have been considered identical.

It is a comparatively recent rediscovery, based on new and sometimes embarrassing scientific facts, that even though the social ideal holds its importance, there is something infinitely more important, all things considered. That is, learning to live in harmony with nature and the natural world.

"But that is extremely primitive!" some may say. "That was the big problem for the cave man, or even for the jungle beast!"

True. It is almost amusing to realize that when the human race was in its infancy, this type of education—this learning to live in harmony with nature—was virtually the only important branch of learning. At first man was strictly an individualist. It was every man for himself, with no law except that of the jungle. If he could make a living by co-operating with nature, he had fulfilled the first law of physical life. That, for his mores, was enough. The remarkable thing is that while he was in this primitive state he did no violence to the natural world. Such violence, ironically, came only when he became "civilized."

When he became slightly civilized, he discovered the value of co-operation and the community system. This was a tremendously important discovery—in fact it might be called the secret of our intellectual, moral, religious, ethical, scientific, economic, political, and esthetic advancement. Of course nothing in this book is to be taken as subverting this important truth.

Yet here is the curious and unaccountable fact:

As man marched ahead through the processes of social and intellectual evolution, he imperceptibly but certainly and correspondingly lost his ability to get along with nature. The result is that modern man knows a good deal less about ecology than the aboriginal Indians whom he found in this country. As a matter of fact, the chief grievance of most of the thoughtful Indians, from the time of white settlement until the present day, is that "civilized" men slaughtered fish, buffalo, and all other kinds of game so fast that they were inevitably to disappear—that he even used dynamite in his selfish gratification of the desire to kill—that he polluted the streams and tore up the virgin land and despoiled the soil and showed an almost complete ignorance of how to make a living from nature's own great bounty over a period of centuries.

An increasingly large area of the earth's surface is now embellished with man's bleak monuments to his own ignorance. Gobi was once a fertile land, and now it is a desert. Baluchistan, Assyria, Babylonia, Persia, Mesopotamia, Tunisia, Palestine—all these were once garden spots. Poseidonia, in Italy, was a noble city of Magna Graecia, with great temples to Zeus and the other gods. After man applied the plow, silt filled up the outlet of a near-by river and formed an uninhabitable swamp. Utica, in northern Tunisia, was once a flourishing seaport, but the Medjerda River became silted in like manner, and Utica was likewise rendered uninhabitable, becoming one of the ghost cities.

To quote from a wise observer, "The pressure of Rome for wine, oil, and flour from North Africa was a strong factor in converting that once fertile region into the goat desert that it is today."

According to H. H. Bennett, chief of the United States Soil Conservation Service, we have already allowed one-third of all our best topsoil—our greatest and our only indispensable mineral treasure—to be washed into the ocean. Our great forests have been gutted and our metal deposits are shrinking at a terrific rate. At the present pace we shall soon become one of the "have-not" nations. This fact is not well advertised, for our politicians are still patting us on the back for being "the wealthiest nation on earth."

The human race may well be on the threshold of an almost revolutionary development—one which may wrench and shock our habit-thinking concepts into a bewildering array of new designs for living. The new ecological patterns may well prove to be far more important than any conceivable social revolution or global design.

The importance of the produce of the natural world, irrespective of any and all financial expedients or money economy of any type, was poignantly to be seen at a time when the United States was making an effort to become, not only the arsenal, but also the implement factory and grocery store for democracy and the globe in general.

If this country plans to become the rich uncle of all the rest of the world, contrary to the advice of our rapidly shrinking inventory,[1] it will be especially important to look to the yield from natural resources, and become acquainted with the land of the Pilgrim's pride. Iron, copper, and other metal resources are already within a few years of depletion. The old physical frontier is gone forever, though new industrial and scientific frontiers of synthesis are rich in promise. The cream-skimming era is over. Its rapid processes were deceptive. Our wealth is dazzlingly kinetic rather than potential, because of our fast-moving frontiersmen of soil, industry, science, and learning. We have to take an unaccustomed inventory of what we have left—of elemental things that are independent of money, buildings, factories, skyscrapers, or cultural institutions.

The arrival of new designs for living, based directly upon nature and nature's world, is argued by scientific discoveries that are coming with bewildering rapidity. Despite the vogue of drugstore vitamins, hormones, and the like, we have rapidly learned in recent years that vitamin-generating ultraviolet rays from the sun have a definite therapeutic value. We learn that insanity varies with density of population. We learn that the birth rate in large cities is suicidal, hence we are compelled to assume that if the city were taken as the norm, the

[1] *United States News.*

norm would also consist of an extinct human race. We learn that when a geneticist sets out to develop a superior type of corn or swine, he does not begin with a huge, sleek, man-propagated product, but with a primitive nubbin, or the razor-back hog of the backwoods, to retrieve certain qualities that have been bred out of the modern animal by generations of artificially stimulated characteristics.

In the iconoclastic book, *Plowman's Folly*, which has attracted unusual attention, Edward H. Faulkner boldly asserts that plowing is a mistake, even though the plow has been used four thousand years; and that nature's way of producing crops is superior. Although his point of view has appeared extreme to many, he has produced evidence stout enough to prove certain basic assertions.

In the matter of soil conservation, which is vital to our national economy and to human welfare, the technicians have made a comparatively recent, yet perfectly obvious and simple discovery: namely, that nature's own way of providing vegetative covering to prevent soil erosion is far more important than mechanical means like terracing or contour farming. Again we may revert to the essay on "Bluegrass" by John J. Ingalls, written seventy years ago, in which he said:

> Grass is the forgiveness of nature—her constant benediction. . . . Forests decay, harvests perish, flowers vanish, but grass is immortal. . . . Its tenacious fibers hold the earth in its place, and prevent its soluble components from washing into the wasting sea. It invades the solitudes of deserts, climbs the inaccessible slopes and forbidden pinnacles of mountains, modifies climate and determines the history, character and destiny of nations. . . . It yields no fruit in earth or air and yet should its harvest fail for a single year, famine would depopulate the world.

Can we wrench our thinking into the realization that learning to live with nature is even more important in the long run than learning to live with each other? Are we big enough to contain both the wisdom of the primitive man and the wisdom of modern artifice? This is a breath-taking challenge, whose implications are but dimly visible, yet the recent wastage and ruthless destruction of some of our most important natural values provides a vivid lesson. We Americans, like billions of people before us, now realize that we have skimmed the cream, and we have indulged freely in the joys we have artificially created, and even at this early date in our precocious national existence, we face widespread deterioration and degeneration, with human erosion keeping pace with the erosion of our pregnant fields. Because of our enormously speeded-up scientific and mechanical processes, we are racing through a destructive stage that required many centuries in such civilizations as those of Egypt and Greece.

For centuries man has smugly boasted about "conquering the wilderness." No more erroneous or short-visioned concept was ever foisted upon humanity. When man tries to fight nature, he invariably loses. Nature inevitably wins. It is only when man is wise enough to live with nature that he really gets anywhere, regardless of any spectacular social gains he may seem to have achieved.

There is nothing more terrifyingly inexorable than the wilderness which silently, imperceptibly creeps back against man's presumptuous but puny projects. It is when man has most exuberantly used his favorite instruments—the plow and the axe, dynamite, the spade and the pick—that he most rapidly and surely guarantees that the desert will march back against him and leer at him over the ghastly remnants of what he calls civilization. He may build dams to create vast artificial lakes, but if he fails to co-operate with nature and treat the watershed properly, she immediately sets to work to fill them up with silt, washed off the land because of the plow. Even as he gloats over his power, manifested in great batteries of hydroelectric turbines, nature has already defeated him, though he may not know it. It is because he has tried to defeat the wilderness.

Probably most of us agree that the natural world provides the norm for all organic creation. If organic creation, by intervention of man, is forced to depart from it, there are penalties—sometimes severe or even cataclysmic. Is there good reason for believing that nature has given man a special dispensation and proposes to let him set up his own artificial and arbitrary norms for living?

2. *The Ecology of City and Country*

PAUL B. SEARS

MAN is a newcomer into a world that is old. We have no means of getting at very accurate figures, but there is a reasonable agreement about the orders of magnitude which are involved. Since it has become possible to make calculations with the aid of radioactive rocks, no one is likely to meet serious criticism who says that the planet Earth on which we live has had a separate career of something like two billion years. During the second billion years, there was a magnificent procession of animal life but no human beings until the latter portion of the record. Only then do animals appear that were close to man in their physical characteristics.

The irrepressible but well-informed Dr. Hooton of Harvard tells us that man was certainly in existence a million years ago, as evidenced by stone tools and skeletal remains. He also believes that animals entitled to the name "human" were present some millions of years before that, during the Pliocene. Here again the proof is in the form of crudely worked stone, and Dr. Hooton's belief is shared by other scholars in the field. Stretching our antiquity generously and saying that we go back five millions of years, the earth-time that elapsed before our appearance was roughly four hundred times as long as the time we have been here. Four hundred to one is the ratio.

This means that the earth, and the life on it, were well set in their way before we came. The whole system was highly organized, and we, of course, have been the beneficiaries of that organization. Our bodies and our basic behavior are fitted to it, and not to some other kind of system. The grasses and other plants which serve us so well, and the ruminant animals which we find so useful, had been here long before our arrival; had we preceded them, our affairs might have taken a very different turn, and the going might have been difficult. What we know of Mesozoic organisms (little enough to be sure) does not suggest easy harvests or companionable domestic animals.

More to the point, perhaps, is the fact that our long prehuman apprenticeship did not merely lead to the exceptional and interesting qualities which we now have, as human beings. It also fixed into our bodies—skeleton, nerves and muscles, sense organs, and glands—certain fundamental animal characteristics which we cannot neglect if we wish to survive. Our novel abilities are built upon older ones as a foundation. We have emerged, but we cannot escape; and solid sense suggests that, whatever we try to do, we had better see that our most deeply rooted requirements are met as well as we can meet them. Great endeavors, from the Promised Land to Rural Resettlement, have failed, not from lack of the higher human attributes, but because these attributes were not used to establish a sound physical basis for meeting our animal needs. Spirit may rise above the flesh, but like an airplane, it must come down for refueling.

These physical requirements of ours are part of a larger pattern embracing all living things, interlocking them with each other and with the earth and its atmosphere. Life is maintained by the energy of the sun, stored in food compounds by the action of green plants, and thereafter available to sustain life. Less than 2 per cent of the energy which reaches the earth is thus fixed, and the volume of life that can be sustained by it depends upon the degree of organization present—just as the efficiency of an industrial plant in using its fuel depends upon machine design and plant layout.

Such efficiency among organisms applies to the use of materials as well as of energy. The stuff that life is made of must be used and re-used by succeeding generations and shared among many forms of life. Within twenty-four hours one frequently eats materials from the bodies of four or five kinds of animals and more than a dozen kinds of plants, and not as chop-suey either! All of this represents a loan, only, and most of it is used as fuel for the work of a day or so at the longest.

We shall see later that under "natural" conditions this chain of transformations tends to become very close-knit and economical. The system maintains itself. But when man takes over, the system is disrupted. Too often he sustains himself by mortgaging the future instead of by maintaining a true economy. He can only avoid trouble by doing his best to understand the inexorable conditions of the world into which he is born and to act in accordance with that understanding.

Our fathers fitted their actions to the idea of a flat earth across which the warming sun moved daily. Far above the earth was a Heaven with its benevolent but jealous God, and below it an Evil Spirit lurking in his intolerable Hell. For better or worse, we know that the situation is not so simple. Whatever one may think of the reality of good and evil, we have to work out our moral destinies by using impersonal forces whose operation, mercifully, we are beginning to understand.

Our earth is not flat, but round, and without compartments. It is one magnificent landscape which includes us all. It is a realm of law, if not always of order. It is an expression of restless energy, working through countless changes of material, through building up and disintegration and rebuilding of form in infinite variety. It is as full of perils and challenges and possibilities as the sea when man first beheld it.

Man has conquered the sea by learning to live with it—

not by violence and self-will, but by patient wisdom in shaping vessels to meet the waters and ride them, and by skill in operating those vessels and guiding them. Perhaps the sea, which so quickly engulfs our failures to deal with it, is a better tutor than the land, which protracts its penalties through years, even generations. To survive on either, we have to know what we are about.

We proceed on the tenable principle that man is an animal and as such sustains certain relations to his environment. There is no better way to determine the character of a given area than by the kind of vegetation and animal life which it produces if left undisturbed. Most kinds of animals are limited to particular types of vegetation, which in their turn express particular conditions of soil or climate, or both. But man has transcended these usual limitations of a species, so that it should suffice to speak of the more general relationships of animals to their surroundings.

Animals are not able to build foodstuffs out of minerals, air, and water. For this they are ultimately dependent upon green plants; both the amount and the quality of food available to support animal life rest upon the fitness of environment to sustain vegetation. For example, wherever unfavorable mineral supply impoverishes the growth of plants, animal nutrition is correspondingly restricted. Thus a primary act in human welfare is to safeguard everything that is needed to insure the thriftiest possible production of all the varied kinds of plant life upon which man depends. These are by no means limited to his conventional crop plants. Particularly serious troubles arise from disrupted water economy, mineral exhaustion, and soil erosion—all of which may follow the heedless destruction of native vegetation.

The relation between plants and animals is not wholly one-sided, for during the course of their evolution, plants have become dependent upon animals in many respects. Great benefits in the production of clover seed, apples, and other crops result from the presence of a sufficient number of bees. Farmers estimate that the value of quail in controlling insect pests is at least ten dollars a head—one reason why they oppose the unrestricted slaughter of these handsome birds. Thus, aside from their direct use to him in varying his diet, man often serves his own welfare by encouraging the survival of native animals.

It is estimated that there are more than a million species of animal life on earth. Among them are forms occupying every conceivable niche in the economy of nature. The largest single group is that of the insects, in many ways indispensable to man, but also his most threatening rivals for the food supply of the world. They represent the highest form of invertebrate life and prevail by virtue of instinct and fecundity. Against these qualities we have to pit our intelligence. The warfare with destructive insects is endless—against termites, apple worms, houseflies, grain weevils, clothes moths, and leaf-eaters, to name but a few.

In this battle, such means as poisons are costly expedients. Any ultimately satisfactory control has to be more profoundly conceived. The whole human environment has to be so planned and managed that it does not favor the undue increase of insects (and other animal life) which are inimical to man. Examine the average pantry or farmstead or tenement house and see what reckless encouragement to weevils, flies, and rats is there, and you will understand how far we fall short of intelligent control.

As the insects represent the peak of instinctive behavior, the vertebrate mammals, which include man, exemplify the perfection of learned behavior. This is our trump card, as the most intelligent of mammals, and we shall have more to say of it later. But it will do no harm to emphasize the fact here, and point out that nothing, absolutely nothing, is more important to our biological survival than proper training of the young. Before we dare to think of training and education as means of the adornment of life, we are obliged to see it as a biological need.

Coupled with this matter of learned behavior is another quality which distinguishes the vertebrates from the insects. The latter rely upon their immense capacity for reproduction. But in the higher animals, as we pass from the fish through the amphibia to the reptiles and on up the scale to birds and mammals, the whole trend is toward increasing care and protection of fewer offspring. Fish are much like the insects, in laying countless eggs and leaving further events to chance. Among the reptiles there are measures of protection for eggs and even for the young. The mammals protect the developing eggs within the warm body cavity, and guard the young until they can look out for themselves. This tendency culminates in man, where technical infancy is being increasingly prolonged, to the point where few professional men have independent adult status before the age of twenty-five. The idea of fewer, better-nurtured individuals is no sudden dream of Margaret Sanger, but a very ancient and realistic idea of Mother Nature.

Through an unhappy emphasis on the ideas of Charles Darwin—one which he certainly did not intend—we hear much of the red claw of nature, and of dog eat dog. The inference is that ruthless competition is "the" law of nature among animals and that man has this heritage to overcome without help from any "natural" principles to counter it. The truth is that, within an animal species, and often between different species, there is a countering principle of mutual aid and collaboration. It is in this that Count Kropotkin saw a biological basis for human ethics, which thus does not have to struggle along against the dead weight of our animal inheritance. For a long time this idea was regarded coldly, because the noble Count was not by profession a zoologist, but his notions now have support in the form of technical evidence from more orthodox members of the priesthood. (Being good is no longer such a lonely job for man, nor necessarily a monopoly for the saints.) Plato recognized that to make a good man, you had first to make him a good animal.

While the details of man's rise to his present dignity are vague, they are not wholly obscured, and we can see the end product for ourselves. Probably the thumb and grasping hand preceded the graduation of forelegs into arms. Somehow pelvis, legs, and spine became adapted to the strains of erect posture, completely freeing the hands, and giving the vertical, rotating head a new command of all it surveyed. Broad-nailed fingers and the tools they fashioned relieved the teeth of some of their grosser duties and lessened the need for massive jaw muscles.

It is, I believe, a principle of neurology that whatever we do induces growth-changes in our nervous system. These changes facilitate the process of doing the same thing more readily the next time. At any rate, the sheer mechanical possibilities of human activity were enormously multiplied by the changes we have sketched, and the central nervous system became more powerfully organized to keep pace. The human brain developed, and human intelligence emerged with it.

Then, too, the human throat became easily the most astonishing instrument for the production of sound that exists in nature. As the central nervous system kept abreast of the growing complexity of experience, the variety of sound signals did likewise, and language was born, bringing with it an important economy of effort. This innovation in its turn has affected the development of body and brain.

By means of speech, experience could be symbolized, communicated, and handed down. Thus two ingrained characteristics of the higher animals could be tremendously amplified. Training of the young needed to rely no longer so heavily upon imitation and crude compulsions. Collaboration could be based more firmly than ever upon mutual understanding and common purpose. The simple patterns of animal behavior were thus transformed into something so much more powerful, responsive, and resourceful as to seem absolutely new and independent of its earthy origins.

This something is human culture. It comprises the whole of man's arts of subsistence, communication, and spontaneous expression. It is the body of his behavior, substantial, enduring, yet malleable beyond belief, changing as he moves from place to place, as he encounters new conditions of living and new groups of his own kind, as he thinks new thoughts and translates them into speech and action.

Except for the rhythmic functioning of his bodily organs and his immediate nervous responses, it is through his culture that the individual human being is related to the world around him. Culture is in a sense intangible, in another sense as real as a running stream. It suggests a musical composition, whose form only becomes obvious in the action of playing, although it can be recorded in descriptive symbols.

The culture of a particular group establishes the relations among its members. It determines what resources in the environment will be drawn upon and how they will be used. Shaped to fit the environment, it serves to mold that environment for better or for worse, and by suitable changes, enables man to enter new and strange environments. For most animals, as for most plants, there exist rather narrow limits of temperature, moisture, and other conditions within which a given organism will survive. By means of his culture, man transcends the limits of climate and locale which restrict other forms of life. Today man covers the earth.

This condition is without known precedent. It is evolution without corresponding change in bodily form. The culture of the Eskimo differs far more from that of the Algerian than the behavior of two species of cat, wolf, or antelope might differ in nature. So far as we know, until man appeared, living things usually had to change their bodily character to meet new conditions, or else perish. For the first time in history of the earth a species—man—possesses keys to its own future evolution, through conscious control of behavior, using the body it has. At any rate, that body is good enough for a better future.

Man is biologically unique in another respect. For the first time—so far as we know—a single species has not only spread over the earth, but it has become the dominant form of life wherever it is found. Our responsibility now has two facets—we are custodians of ourselves and of our environment as well. We did not make and cannot change the laws under which we must work, but at least we can understand them. That understanding is one of the most serious duties of our culture.

Throughout most of the putative five million years of human or humanoid existence, the resources of man's culture have been taxed merely to keep him alive—to serve the primitive animal needs of physical survival. So long as he remained a gatherer, a hunter, and fisher, there was little margin of time or energy for anything but subsistence. Men had to spread out thinly, in small groups, for this sort of existence requires space. Then, at a time we can place roughly at ten thousand years ago, there arose the art of agriculture, based upon the domestication of plants and animals. By affording abundant food with less labor and space than had been required under earlier ways of living, this art gave men some leisure and permitted them to live in larger groups than before. Such were the modest beginnings of what we call civilization. Referring again to our round numbers, let us get the whole situation into perspective:

$$\frac{\text{Age of Earth}}{\text{Age of Man}} \qquad \frac{2{,}000{,}000{,}000}{5{,}000{,}000} \qquad \frac{400}{1}$$

$$\frac{\text{Age of Man}}{\text{Age of Man since civilization began}} \qquad \frac{5{,}000{,}000}{10{,}000} \qquad \frac{500}{1}$$

$$\frac{\text{Age of Earth}}{\text{Age of Civilization}} \qquad \frac{2{,}000{,}000{,}000}{10{,}000} \qquad \frac{200{,}000}{1}$$

Relatively speaking, civilization is not a very old experiment, yet it must deal with forces of impressive antiquity.

Just what is civilization? It can be looked on in matter-of-fact fashion as an elaborate technique which enables man to survive in highly concentrated populations. This really puts it on the same moral level as more primitive culture, making it minister to the old animal necessities as a chief end.

Or civilization can be construed in the words of Matthew Arnold as "the humanization of man in society." This is my view of it—the use of leisure and all other advantages to secure the physical needs of course, but doing so as a means of fostering whatever is distinctly human, that is humane. This then, is the test by which we ought to judge our modern culture. There are three useful words to be pondered in this connection—citizen, civic, and city.

The citizen is the individual participant. The title implies a traditional dignity, worth, responsibility. Civic enterprise implies co-operative, rather than competitive, effort towards the ends of civilization, that is towards the encouragement and conservation of humane values. The city is the locus of group activity, a place where men dwell and work in numbers, by grace of food and other materials produced elsewhere. It is an instrument, and unless it contributes to civilization as we have defined it, it has no particular justification. If it fails, as it often does, not only to foster what is human, but even to insure adequate care of the animal requirements upon which humanity rests, it becomes a sorry institution. In fact, what we call "common humanity" is our concern to prevent needless physical suffering on the part of those less fortunate than ourselves.

Just what is meant by the term humanization? Roughly the word is emphasized by its antonym—brutalization, but I prefer to be more concrete. Football, for example, may be either civilizing or brutalizing. The clue lies not in a particular activity but in the end towards which it is directed, and the way—the spirit—in which it is conducted.

Adequate nutrition and the chance for physical development have to be taken for granted as the basis of humanization. Next I should place training and use of the remarkable bodily features by which man is distinguished from other animals and which have already been mentioned—grasping hand, erect posture, radial vision, and highly developed central nervous system. In all seriousness, when a man neglects to cultivate his ability to walk erect on his hind legs, or simply to discipline his central nervous system (whether he will employ the resultant skill in earning a living or not), he is to that extent dehumanizing himself. If the use of his automobile encourages atrophy of leg muscles, and dependence upon an adding machine discourages reasonable facility in calculation, these inventions are not, in those respects, instruments of civilization.

Over and above such rudimentary physical matters are the obligations of civilized man to employ his culture towards co-operative, rather than ruthlessly competitive, action, and towards the conditioning of his young for a kind of life that will further human—that is to say, humane and moral—ends.

Thus civilization, in the only sense that satisfies me, means a consciousness of the role of culture and a sense of responsible participation in shaping and maintaining it. For assuredly it is man's most unique biological attribute. Call it spiritual if you will—the obligation exists in any case.

In biological accounting, the ledger record of any part of the body, or any organism in the community, must balance reasonably well or be closed out by failure. If a group of cells in the body "go wild" and grow at the expense of useful cells, without making any contribution to the economy of the body as a whole, we have malignant disease. The bacteria which feed upon the roots of clover also fix nitrogen from the air—more valuable to the clover plant than the sugar which it furnishes in return. Predators, such as wolves, seize first the defective young of deer and other animals upon which they feed. In the absence of predators, it has been shown that the quality of deer is likely to depreciate. But if the wolves are too numerous, they may consume healthy breeding stock as well, and the number of food animals may decline to the point where the wolf packs themselves are starved out. The idea of balance—of a flexible system of give-and-take—seems implicit throughout nature. It is the basis of physical and chemical theory. It is fundamental in biology. There is no reason to think that human activities are exempt from its control.

Certainly it operates in the growth and distribution of population, and hence in the development of cities. The capacity of the earth to support a human population varies from place to place, but under primitive economy it is quite restricted. Under such conditions infant mortality is usually high, abetted by contraception, abortion, and infant exposure. The distribution of food under such systems simply does not permit the accumulation of large numbers of people within a limited space. Pressure develops quickly—doubtless one reason why the human race spread so widely over the earth before it became civilized.

The state of Ohio, which has at present a population of seven millions in an area of forty thousand square miles, is believed to have had not more than twelve thousand Indian population just before white settlement. Under the primitive economy of hunting, fishing, and some simple agriculture, there were over three square miles, or about 2,100 acres per person. There was not much slack in the pattern of thinly distributed aboriginal population in North America. For such a way of life, much room was needed, and we know that the early European settlements on the Atlantic Coast caused a wave of profound disturbance that moved far into the interior. Under our modern industrial economy there are 175 persons per square mile, or a little less than 4 acres of land per capita. Even so, the population is much less concentrated than in Massachusetts, Belgium, or the Nile Valley, where it ranges between 800 and 1,000 per square mile.

Relief from such intolerable pressure of population can only come through the classical checks of war, disease, and starvation, or through migration, unless the birth rate is controlled or a new subsistence relation to the land is developed. When an efficient agriculture is established, food can be produced with less labor, in less space, and in much greater amounts. Population can concentrate in centers, and a growing number of individuals can be released from the production of food to attend to other matters.

The first use of leisure contributes directly to survival—building, shaping, defending, distributing, and governing. But quite as insistent seems to be the need for expressing the peculiar character of the group and establishing its forms of behavior. A spear-shaft or a garment unadorned is just as useful as one with ornament, yet the demand for pattern and color is all but as imperative as the need for utility. The singer or story-teller is cheerfully fed by those who till the soil, and so are the artist and priest.

It is at this level that qualities and dilemmas peculiar to civilization emerge. The Spartans, like the modern Germans and Japanese, exalted the soldier above all other elements in the population. The Chinese ranked the scholar first, the farmer next, with artificer, merchant, and soldier in descending order of esteem. Civilizations, like individuals, are to be judged by what they do with their leisure. They are humanized or brutalized by the way in which they employ the margin of energy left over from providing for the physical necessities.

More immediately, the form of living that is developed by this margin of leisure, the values that arise from it, reflect back to the work of those who produce food and provide the basis for the whole enterprise. Men strive harder for esteem and prestige than for any other rewards. The farmer does not like to be meagerly rewarded, but he minds a great deal more when he and his children are looked down upon by those for whom he provides. Where the land and those who work upon it are forgotten or despised, the very existence of the urban centers which develop this attitude is speedily threatened and must ultimately give way.

The later history of the Roman Republic, for example, can be read largely as a struggle for dignity and recognition by those who worked the land against those who owned it and who lived in the great centers of population. Back of the contests of the Gracchi and the tragic career of Julius Caesar lay this problem. During the youthful period of any civilization, the men who till the land are not ashamed of their vocation, but the trend is almost always towards concentration of ownership in a few hands and a lowered status for the man who works the soil.

Much of the stability of China is due to its rational scale of values. In the opinion of Ralph Linton, China has maintained a relatively higher level of civilization for a longer continuous period of time than any other human empire. In contrast, whatever else contributed to the fall of Roman civilization, the failure involved an unsound and unrealistic attitude toward the problem of subsistence. The relation between cities and the rest of the landscape came to be not one of mutual obligation and benefit but rather that of parasite and host.

It is an established principle of modern medical science that you cannot deal with one part of the human body and ignore what happens to the rest. Even when a tourniquet is applied to stop the bleeding in an arm or leg, the bandage must be loosened at intervals to restore circulation. It is likewise a principle of modern physical science that the whole universe is interrelated in terms of cause and effect. To use a more modern phrase, it is all part of one great intermeshing web of probability.

Thou canst not pluck a flower
Without troubling some bright star.

It is elementary science and not mere analogy to say that the landscape occupied by a human culture cannot be divided into parts isolated from each other, nor can it remain isolated from the other cultural landscapes of the earth. There have been many attempts to place areas in quarantine, but they have never succeeded. The long political struggle for which William Jennings Bryan served as the spearhead in 1896, and which is still going on, resulted from the attempt of the industrial East to profit at the expense of the agricultural South and West without concerning itself with the effect upon the lives and welfare of this rural portion of the landscape.

Not only is the landscape a unit, but it has the quality of health or disease. This again is not an analogy. Health is a condition of normal functioning, disease a disturbance of such normal functioning. In a normal or healthy landscape, as in a chemical reaction, the trend is always toward equilibrium or balance. The processes, while at times violent, tend to be regulated and orderly. This is well exemplified by the behavior of water upon whose endless cycle of evaporation, movement, precipitation, and slow return to the sea all life is dependent. In an ordered landscape, the water which falls and does not return to the air either soaks into the ground or moves through organized and balanced systems of drainage downward and off the land surface. The action of falling water upon the earth is cushioned by the carpet of vegetation, which regulates its violence and thus makes possible the deliberate use of water by living things before its ultimate return to the clouds or the sea. Water soaking into the earth is stored underground to be fed back to the vegetation and animals as soil moisture or in the form of springs. Destructive flow of surface water is held in check by vegetation and by the ordered gradient of natural drainage systems. This is important, for whenever the velocity of water is doubled, its capacity for mechanical action may be increased as much as sixteen-fold.

On such a landscape, too, the processes of soil formation go on without substantial interruption. Physical and biological changes produced by plants and animals tend to make the soil more livable because it has been lived in. Nutrient substances are mobilized in it, its relation to water, air, and life is tempered, so that it tends to become as productive as climate and other conditions permit. This kind of order is expressed in the richness, variety, and efficiency of plant and animal life. The sun's energy, falling upon such a landscape, tends to be used with as little waste as possible in the form of radiated heat.

One has only to compare such a landscape with a flat area of newly exposed rock or sand to appreciate what is meant by its equilibrium and its unity. The forces of sun and atmosphere strike such a rock surface and act violently upon it. The rain runs off without check or gathers into depressions. It may run down through cracks but there is no sponge-like structure to retain it for the use of plants. Without vegetation there is little or no animal life. Heat and cold, wet and dry—these extremes are typical of such an area.

Of course in an economy of nature undisturbed an area of this kind does not long remain barren and inefficient, for it is always part of the larger landscape. The moment it is exposed to the forces of the atmosphere, the slow processes of erosion and soil formation which lead to eventual equilibrium are inaugurated. Left to themselves, these processes will bring the barest and most unpromising surfaces to a condition of productiveness and order. It is true that any stage in the process of this course may be interrupted or slowed down by such things as volcanic action, changes in sea level, or shifts in climate. Yet so long as the surface remains exposed to the air, such catastrophes merely delay but do not end the slow swing towards balance and order.

On the other hand, since man has become so numerous and has such powerful techniques at his disposal, he has been able to disrupt this natural economy of the landscape without always substituting an economy consistent with the same great principles. He removes vegetation and substitutes crops which leave the ground bare for a good part of the year. He constructs systems of highways which are in effect artificial pathways for the drainage of water. Too often, especially during the last century, these highways have made little compromise with the natural topography, speeding water on its way, diverting it from the upper reaches of the original drainage systems and dumping it in floods into the larger streams. Under this scheme not only does the water fail to soak into the land so that it can be used to sustain life, but it carries with it an increasing freight of the slowly built and fertile topsoil. Men themselves are crowded into cities, not only losing touch with the rural landscape which feeds them, but lacking the elementary provisions for normal physical exercise, balanced diet, and fresh air which are necessary to maintain the human animal in good physical condition. Under these conditions, the trend of the landscape towards order and equilibrium is not only arrested but often reversed.

It is characteristic of such a landscape that it is seldom esthetically satisfying. It is no longer beautiful nor healthy. Its ugliness serves as a symptom of its disorder.

With the discovery of the method of invention and the tapping of great natural forces to do the work which men had done by hand, men's capacity to disturb the landscape has been tremendously enhanced. This is sometimes interpreted to mean that science is a monster which may turn on man and destroy him. Such is not necessarily the case, for the knowledge of natural forces can be utilized to promote a new equilibrium which will make the landscape efficient.

If we examine the application of science in modern industrial civilization, we find that with the exception of medicine and agriculture, the application has been almost exclusively at one stage in the process. This is the elaboration of raw materials into consumers' goods which can be sold at a profit. This phase of civilized activity is emphasized almost to the exclusion of two other steps equally vital: (1) conserving the supplies of raw materials; and (2) intelligent distribution on the basis of need. It has sometimes been said that modern industry is largely engaged in designing and making things and then persuading people that they ought to have them.

Such a use of science ignores any principle of equilibrium. It fosters a neurotic civilization and focuses attention on things which are mere instruments, rather than on the quality of living and the main business of civilization which is the humanization of man.

Such a use of science obscures its most important function, which is that of analyzing our relation to the landscape and to each other in order to get a check on the direction in which we are moving. Every great religion has taken cognizance of the fact that the behavior of human beings is ultimately shaped by the kind of world they *think* they are living in. In so far as science affords us new concepts of our world, it has the capacity to modify our behavior, and has done so to a remarkable degree. But where civilization employs science primarily as an instrument for private profit (or warfare) there is no guarantee that the average individual will have a much sounder perspective of the world than he had in the days before science. In fact his view is likely to be more confused, less integrated, less serviceable even in the relationships of everyday life. The citizen of the new technical world has had an older, simpler world-picture shattered to pieces, with nothing very satisfactory, as yet, to take its place. Meanwhile he is kept too busy, by the rapid tempo of an applied science, to do very much about it—in the way of quiet thought.

The sovereign function of science is not to save us from muscular or mental exertion; we need both for the health of mind and body. Rather it is to push back, so far as possible, the bounds of uncertainty which press on us from every side, to give us such confidence as we may command, to restrict disputes and uncertainties that are needless. It is no function of science to release us from obligation, but rather to show where obligations lie and how uncompromising their claim is upon us. There is nothing fundamentally wrong with making men more comfortable, or their living more convenient, but it is decidedly a secondary order of business.

Consider for a moment the way in which our marvelous knowledge of biological science has been used. True, it has been applied to the art of agriculture. We know infinitely more about better varieties of crops, about pests and their control, about plant nutrition and the care of farm animals than did our grandfathers. Yet the yields in our best farm states have not been commensurate with this increased understanding. In our concern for immediate profits, we have neglected the soil, which is capital wealth, and allowed its productivity to decrease through erosion and other forms of exploitation.

Great progress has been made in medical art, too, yet the insistent pressure here has been to arrest the epidemics which threaten great industrial concentrations of people. The cultivation of general physical fitness through proper diet and exercise has lagged far behind. Much has been done to secure adequate knowledge of diet and infant care for the poor, yet the funds (which are a measure of social energy and intent) available for that purpose are meager in comparison with those available for the spectacles of sport, where the vigorous few are employed to thrill the flabby multitudes.

It can be argued that the application of engineering technology for the profit motive has not worked too badly. The techniques of mass production and advertising have certainly bettered the standards of living in the United States. Yet the progress has not been sufficiently uniform or satisfactory to bring about even a reasonable measure of social and industrial peace. Among other nations, less blessed with resources than are we, the effect of science on living conditions for the general population has been so unsatisfactory that the spirit of aggression has been easily aroused, drawing the world into a second general war. The advent of war, instead of finding us alert and realistic, as becomes a people trained to face hard facts, found us at first divided and undisciplined. Only the shock of a violent and overwhelming surprise attack served to arouse us and polarize our energies. Such was the price of our moral confusion, in the ugly twilight between the decline of our old faiths and the birth of a new and general enlightenment.

If our new technical world has not brought us moral firmness, it has other defects as well. On the plane of our physical requirements it has failed us, not only through pre-occupation with profits and inattention to the common good, but by promoting activities subversive to our simple bodily welfare. The struggle to prevent the contamination and adulteration of commercial foods, launched so dramatically with the "embalmed beef" scandal of the Spanish-American War, has been unbelievably difficult. Beside the tough fighting spirit of Harvey Wiley, constant vigilance on the part of civic groups and their government servants have been required to win this battle and keep it won.

The story of wheat-flour is by now generally known. During a long generation, following the perfection of a process for milling refined white flour, we paid with defective teeth and lowered vitality for a product which pleased the eye and could be kept indefinitely for shipment and storage. Removal of the rich germ removed the oil which renders old flour likely to become rancid. But in that germ lie vitamins essential to human health, especially important to urban populations far removed from the abundant crude fresh foods of the farm. Fortunately the solution is fairly simple. Wheat itself keeps admirably, and small mills for grinding it fresh near the point of consumption are entirely feasible. It remains doubtful whether we shall get them through public insistence. The more likely course is that some well-capitalized concern will risk the wrath of established industry, start making the necessary small household mills, and then expend huge advertising sums to convince us that we need these devices for our health's sake. Eye appeal in the mechanism will not be neglected, and we shall doubtless be taught that the "smart hostess" grinds her own, preferably while clad in girdle and garters.

Even at its source in the fields our food is not exempt from the pressure of an unbalanced scientificism. The mass production of cash crops has received far more attention than the establishment of a diversified agriculture which will maintain and enhance the quality of the soil, and thus of rural and urban diet. Most of our cotton is grown on land dangerously low in mineral content—a condition made possible by the fact that cotton fiber is a carbohydrate which does not require much nitrogen, phosphorous, or potash in its composition. Grim pity that the spirit of young men from the cotton producing areas was not matched by their physical condition! The percentage of rejection for the army in World War II was notably higher for regions of low mineral content in the soil than for areas where the nutrient conditions were naturally better to begin with and had been less impaired by the prevalent agricultural practice.

Much has been made of the cry, "human rights versus property rights," by those who see clearly the evils wrought by an unbalanced application of science. But this is a red herring across the trail. There is no necessary conflict between the two kinds of right. The right to own private property is in itself one of the most valuable and hard-won of human rights, an invincible source of morale, of the feeling of identity and responsibility. The Swiss are on a workable basis for democracy with their feeling that every voter ought to own a bit of the fatherland—even if it is only the site of a cowshed. We shall never reach internal peace in this country by the route of abolishing "property rights," but rather by diffusing them more widely and giving them a greater sense of continuity and permanence.

There is an alternative to the concept of applying science too exclusively for private profit. It lies in more emphasis on the biological process of co-operation, and less on unrestricted competition. Back of all private ownership, even in our own system, lie the principles of escheat—ultimate government ownership—and of eminent domain—the supremacy of public over private interest. Even the most powerful corporations hold their possessions under sufferance. There is nothing un-American in a more widely diffused private ownership than we have today, with consequently more pressure for benefit from science, and less for profit from it. In fact, the idea is basic American doctrine.

The biological influences of the new technical age are not limited to problems of nutrition and care of the soil. They have struck at the family itself. One of the by-products of our situation is a growing tendency to shell off responsibilities to more or less clever experts. We have developed a mood of delegation, which may be justified if we have shoes to be mended or an intricate wiring system to be set right. It may be justified, too, for a considerable part of what we call education, but certainly not for as much of it as we are passing to the schools by default. The core of the job ought to be done in the home, and is only to be shifted elsewhere at the greatest hazard.

The old nomads of the desert moved their homes from place to place, it is true. But what they moved was an organized and coherent institution, in which each individual played a responsible role. We are becoming the new nomads of the scientific age. A home today is a place to eat and sleep, not a common enterprise for those who share it. The functional role of the children has all but vanished, and with it our most rational opportunity to train them at home. Thus to our habit of turning over responsibility to others is added a tangible physical difficulty. Few children are so dull that they cannot sense the difference between "made work" and the real thing. Here again a wider diffusion of private ownership must work toward re-establishing the significance of the home. We hear much of the evils of farm tenancy; but they are far more curable than those of urban tenancy.

If you wish to destroy any kind of animal, you have only to attack its nests. The home is the nest of mankind, and its protection is the paramount duty of any culture. It is reasonable to expect, in a scientific age, that this responsibility would be the first to benefit by scientific enlightenment, and beyond doubt it has done so. Yet one has only to compare the technical energy being devoted to activities which foster self-indulgence, divorce sex from personal loyalties, and generally weaken the prestige of the home and the family, with the efforts being made to conserve these institutions, to see where the emphasis lies. Even those of us who believe in the ultimate power of scientific enlightenment to dignify personality and enhance the honesty and permanence of human relationships can see that this line is mostly being held, not by those who work in the name of science, but by those who adhere to older disciplines.

Concerning war in the new technical age, the failure and tragedy which it represents is obvious.

If we turn now from human affairs to examine natural communities of plants and animals, it is neither as an escape nor a search for romantic analogies. Any community is a system of forces, or processes, and there are fundamental laws which appear to govern such systems, however complex or of whatever order of magnitude they may be. In particular, there is the law of Le Chatelier and Bancroft, which may be stated as follows: "Any system tends to move toward a condition of minimum stress within and without."

If we examine a community getting under way, say in an abandoned field, we are struck with the predominance of vigorous weeds, insects, and casual animal visitors. Relationships are highly competitive, neither mutual, intimate, nor permanent in character. The very weeds are unable to thrive in their own shade, and are replaced quickly by forms which can do so. The forces of nature impinge upon such an area violently and directly, nor is there an efficient use of energy.

With the passing of time, a process of succession and development takes place, moving always in the direction of balanced relationship, effective use of energy, and a kind of dynamic stability, until, under favorable conditions, a community is present which sustains itself without much change in character. Stresses within the community and between the community and its surroundings are at a minimum, and energy is used with as little leakage as possible to maintain the system in a condition of high productivity.

In such a moving equilibrium each organism, plant or animal, has its niche of activity and its peculiar role. The stability is not one of stagnation, but of intense activity. While the character of the milieu is set by the dominant organisms, these are often quite dependent upon the activity of inconspicuous and seemingly subordinate forms. Thus in certain types of forest, the dominant trees could not survive without guest fungi which are present in their roots, serving as intermediaries with the soil. Many animals, clearly dependent upon the plants for their sustenance and shelter, may serve plants in varied ways. Relations tend towards a condition of mutuality, of collaboration and order, in contrast with the untempered competition of the original weedy field.

Not only is energy used effectively in such a community, but there is a favorable energy budget. Each growing season of activity sees more energy impounded as a reserve than is expended for current activities. For we must remember, that whatever else life may be, it is a manifestation of energy. The energy so accumulated takes various forms. It is found not merely as a reserve in the chemical substances which compose plants and animals, but is expressed in the organization of their bodies, and of the soil as well. It is analogous not only to the fuel reserve of a manufacturing plant, but to the energy used in organizing and building that plant.

In such a pattern we have a norm or standard, and not merely an analogy, for the development of man's communities. It suggests the kind of relationships he must use as a model in his own operations on the landscape, in contrast to his corn and cotton fields which so often correspond to a community at the lowest level of organization. The increasing use by the best farmers of perennial pastures and close-planted fields which are kept in cover most of the year shows that the lesson has not been wholly lost.

But most important, this pattern suggests the type of organization of the whole human community if it is to attain to any degree of order, balance, or permanence. In crude physical terms, the energy budget must at least break even, and ought to show a net gain. It has been computed that the landscape of the United States is at present fixing 10×10^{15} calories of energy each year, while its population is expending 6×10^{15} calories of that amount and drawing upon the underground reserves in the form of fuel for an additional 5×10^{15} calories more. Better management is imperative and happily quite possible, although these computations were made before the hideous drainage of the late war had begun.

And finally, the laws of community development suggest that a stable human community must develop collaboration and the chance for wholesome individual activity for people of the most diverse roles. Competition can scarcely, in the nature of things, be eliminated, but at least it need not be weighted so heavily against so many individuals as at present.

3. *It Was Not Always So*

WARREN S. THOMPSON

HISTORICALLY man is a product of the village. Only a very small proportion of men have lived in cities until quite recent times. The chief reasons man grew to be what he is in an agricultural village rather than in the open country or the city are (a) that he needed the protection of his fellows, and (b) that he could not support any considerable population which was not engaged in producing food. If he lived a pastoral life, even the small clan group needed a relatively large area for its support. It was not feasible to have more than a few score persons in any village of nomads. If he tilled the soil, the distance from the village to the land he worked became too great if the village had more than two or three hundred people. This is also confirmed by the fact that in many parts of China and the Soviet Union today the agricultural village does not average over two hundred persons. Since, either as pastoral nomad or as tiller of the soil, the agricultural worker produced very little beyond his own keep, there was almost nothing left to support a nonagricultural population. All over the world until about one hundred and fifty to two hundred years ago, and even now in a large part of it, man had only his hands and a few crude implements and tools with which to make his living. Under such conditions he could spare very little food from the farm.

The proportion of the crop that could be spared from the sustenance of the agricultural worker for the nonagricultural population remained extremely limited until more scientific agriculture and more extensive tillage came into being. But, of course, this proportion varied from place to place according to the fertility of the soil, the favorableness of the climate, the type of crops raised, and finally, but not least, the density of the agricultural population. Under very favorable conditions the agricultural worker in an age of hand production may have been able to spare to the nonagricultural population from one-fifth to one-fourth of what he produced. Under less favorable conditions, often as the consequence of an increase in his own numbers on a limited area of land, he could not spare any of his crop without endangering his own existence.

Twenty to 25 per cent is, of course, only a guess as to the upper limit of the proportion of the crop that could be spared from the land as long as a hand system of agriculture prevailed, but it seems a fairly reasonable guess in view of what we now know about the proportions of rural and urban population in some countries where this system has prevailed until recently and in those in which it still prevails. In Czarist Russia (before 1914) the rural population constituted 85 per cent of the total and even the first Soviet Census in 1926 gave the rural population as over 82 per cent. Of course not all this rural population was agricultural. Some of the people in villages were engaged in making simple tools and implements for the peasants, in trading in farm products, in milling the grain, and in performing various other services useful to the peasants. Even so the proportion of the population actually engaged in agriculture may have been as much as 75 to 80 per cent.[1]

It is quite commonly believed that in China 80 to 85 per cent of the people live in agricultural villages and market

[1] Frank Lorimer estimates 78 per cent primarily dependent on agriculture, forestry, or fishing. ("Population Movement in the Soviet Union," summary of a paper presented December 7, 1943, at a round-table conference under joint auspices of the Research Bureau for Post War Economics, the Russian Economic Institute, and the Committee on Post War Population Resettlement.)

towns and that nearly all of them are farmers. The proportion of actual agricultural workers is probably much the same as in pre-industrial Russia. Chinese rural communities, including the market towns, are still largely self-sufficient. They produce so little above local needs that it is hard to believe that more than 15 to 20 per cent can be sent elsewhere to support people in cities. As regards China, we should not let the rapid growth of great port cities like Shanghai and Tientsin obscure the fact that, although city populations are growing, only a very small proportion of her people even now live outside the agricultural villages and the market towns, and that probably not more than 5 to 7 per cent live in cities of 100,000 or more. Nor can this proportion increase significantly until the per capita production of the Chinese agricultural laborer increases.

Indian agriculture appears to be able to support a somewhat larger nonagricultural population than does Chinese agriculture. The latest data indicate that perhaps 72 to 75 per cent of the population is agricultural.

Even in the United States (see Table 1) we are not far from the time when 80 to 90 per cent of us lived on the land,

TABLE 1: *Distribution of the Population, by Size of Community, United States:* 1790–1940.

Area	1940 *No. of Places*	1940 Per Cent	1900 *No. of Places*	1900 Per Cent	1850 *No. of Places*	1850 Per Cent	1790 *No. of Places*	1790 Per Cent
Total	----	100.0	----	100.0	---	100.0	--	100.0
Urban	3,464	56.5	1,737	39.7	236	15.3	24	5.1
100,000 & over	92	28.8	38	18.8	6	5.0	--	----
25,000–100,000	320	11.2	122	7.3	20	3.8	2	1.6
10,000– 25,000	665	7.6	280	5.7	36	2.4	3	1.2
2,500– 10,000	2,387	8.9	1,297	8.0	174	4.0	19	2.3
Rural	----	43.5	----	60.3	---	84.7	--	94.9

or in the small villages which were farm and service centers for the agricultural community; nor did we at that time ship much agricultural produce abroad. There is no need to labor this point; obviously a hand agriculture and a hand industry with poor or nonexistent transport between communities made it practically impossible to have more than a very small city population. This is the basic reason for our certainty that man has been a rural village dweller during practically all of his existence on earth as an agriculturist.

We are reasonably sure also that until quite recent times most of the cities that did exist were quite small. Even great cities like Rome and Peiping at the height of their power probably did not have as many as a million people, and they grew to this size only because they were the centers of great empires whose rulers could levy tribute on a great agricultural population, so that even a small surplus per capita came to a large absolute amount. Moreover, this surplus could be brought by water to within a few miles of these capitals. In the past, practically all large cities have had fairly good water transport even if, like Peiping, it involved building long canals.

India, unlike Rome and China, never had a highly centralized government, and as late as 1872 when the first census was taken, had only two cities (Calcutta and Bombay) of 600,000–650,000 and only thirteen cities over 100,000 population although the total population was about 256,000,000. In Europe the city of over 500,000 persons is a quite modern development. In 1700 London alone is supposed to have exceeded this number (674,000), although Paris may have been nearly that large. It was not until after 1801 that London passed the one million mark.

Modern urbanization was made possible by the increasing productivity of agricultural labor. Without an agricultural revolution, closely associated with the Industrial Revolution, there could have been no large increase in urban population. In addition, the improved transportation accompanying the Industrial Revolution made it possible to bring this larger agricultural surplus to the city from greater and greater distances at low cost. Thus the agricultural revolution which began in the West about 1700 made possible a great movement of population from agricultural to nonagricultural work, and modern steam transportation made possible the great city of today. There is no need to dwell longer on the release of labor from agriculture made possible by more efficient agriculture, but it will be well to enlarge a little on the effect of steam power upon the location, the size, and the structure of the modern city.

In pre-steam days it was almost impossible to feed any large city which was not readily accessible by water; nor could the few manufactures the peasants could buy be distributed over a wide territory. The cost of carriage by pack horse, wheelbarrow, and shoulder pack was prohibitive beyond the radius of a few miles, except for very expensive goods, for example, spices and silks. Moreover, transport of goods by wheeled vehicles was, with only a few exceptions, impossible before 1700. There could be a certain amount of cartage within the local community, but it was very limited even there.

The advantage of water location was lessened with the coming of the railroad, but it must be remembered that the railroads operating before the middle of the nineteenth century were of almost negligible importance from the standpoint of moving goods, even in England. Most of the great cities of today, even those in the New World, were located in the days when water transport was, relatively, far more important than it is now. When railroads came, they were built and run in such a way as to encourage the growth of the cities already established rather than to encourage the development of new centers largely or wholly dependent on rail transportation. This is true in general, although some fairly large cities have since been built which have no water transport. But steam power not only made large cities possible, it also exercised a decisive influence on their structure and organization.

Steam has to be converted into usable power quite close to the point of generation, and it is most effectively generated in relatively large units. Thus steam as an agent of transportation favors shipping in large quantities and over relatively long distances rather than moving small quantities over short distances with frequent stops. The large city which could use shiploads, trainloads, or at least carloads, had a natural advantage over the small city or town which could only use carloads or broken carlots. This natural advantage of the large city when serviced by steam transport was generally supplemented by rate and by service advantages. Altogether there can be little doubt that steam transport was a potent factor in encouraging the growth of the larger centers already established either as commercial centers or as manufacturing centers. Railway transport also favored the highly centralized city by delivering large numbers of people and large freight consignments to a relatively small downtown area more expeditiously and, up to a certain point, probably more cheaply than smaller numbers and amounts to several scattered areas within the urban region.

The centralizing effect of the use of steam transport on city growth was, however, probably much less than that of its use as power to drive machines. Until about forty years ago power had to be transmitted from the steam engine to the machine by mechanical means—pulleys, shafts, and belts. This made it economical to build the factory around the power plant—no machine could be more than a hundred yards or so from the engine, and the engine had to be quite close to the boiler. Thus the very nature of steam encouraged the building of larger and larger factories around the power plant and concentrated more and more workers near it; for, as we have seen, steam was not well adapted to quick and cheap local transport.

It was not until about 1900 that a new structure of urban life became possible, and it was another two decades before the potentialities of electricity as the agent of a new industrial and social revolution were widely realized, if, indeed, it can be said that they are realized even now. With the development of the use of electricity for power and communication and with the perfecting of the internal combustion engine for automobiles and airplanes a whole new distribution of industry and commerce and, therefore, of population became possible.

Long after it ceased to be necessary and probably after it ceased to be economical to crowd an enlargement of a factory into the small radius within which steam power could be transmitted to the machine by shaft, belt, and pulley, habit and convention continued to dictate the location of new productive equipment. Likewise, long after the need for frequent and prolonged communication between the men managing the production and the commercial aspects of a business ceased to depend on their having offices next door, or in the same building, or even in the same city, the process of agglomeration went on. But even when it was realized that the telegraph and especially the telephone made the decentralization of a business possible, there was little or no effort to use this new power to decentralize population. When the top-flight management took the opportunity to escape from the horrible manufacturing centers it had created, it only left these to go to larger commercial and financial centers. Furthermore, it quite frequently took a large part of the clerical staff to the new home. The net effect of management's moving out of manufacturing centers was not to decrease population in these centers so much as to increase congestion in the great commercial and financial centers. This inevitably resulted in increasing still more the proportion of the nation's commerce and wealth flowing into such centers. There does not appear to have been any significant change in this pattern of urbanization in the United States until after 1930, if then.

The centralization of the higher-flight executives of concerns doing a national business seems to have gone ahead without interruption between 1920 and 1930; indeed, it seems to have been accelerated during this decade. There was a veritable stampede, by the larger concerns doing a national and/or international business, to establish headquarters in New York, Chicago, or some other great city and for concerns doing a regional business to do the same in the dominant city in their region. It is true that after 1920 the large cities themselves did not increase as rapidly as their suburbs, but the decentralizing effect of the increasing use of the auto, the telephone, and electric power did nothing to break up the congestion of the downtown portion of our great cities. All they did was to scatter the living quarters of a growing proportion of city workers over a larger area and to move some of the industry of large cities from downtown locations to peripheral areas.

Sometimes the effect of moving industries to the periphery of a city without making any plans for housing was to make it even more necessary for the workers to live in a central location from which they had ready access to all parts of the periphery. But the lack of planning in connection with the centrifugal movement of industry was only one factor in keeping population from decentralizing to the same degree. Another very important factor was the reluctance of the workers to move to a locality where they could not readily secure transportation to other factories. They do not like to become too dependent upon a single employer or even on a few employers.

Between 1920 and 1930 the ninety-six metropolitan districts of 1930, nearly all of which contained a city of over 100,000, increased over 28 per cent in population, while the rate of growth for the total population was only 16.1 per cent. The central cities in these districts, however, grew by only 22.3 per cent, while the districts outside these cities grew by 44.0 per cent, or practically twice as fast. All the cities not in these metropolitan districts increased by only 19.4 per cent, or at a significantly slower rate than the central cities and only about two-thirds as fast as the entire districts. The rural areas of the nation outside these districts increased by only 3.7 per cent. When the rural population is broken into farm and nonfarm, it is found that the farm population declined by about 1.2 million while the nonfarm population grew by 3.7 million.

Between 1930 and 1940 this trend towards the metropolitan districts continued, although at a slower rate. However, instead of the metropolitan districts growing almost twice as fast as the nation, the 133 metropolitan districts for which comparable data are available grew less than one-seventh faster than the nation (8.2 and 7.2 per cent, respectively), while the districts outside the central cities grew more than three times as fast as the central cities (15.8 and 5.0 per cent, respectively). For the first time the outside areas grew more in absolute numbers than the central cities. It should also be noted that the small cities (2,500–50,000) grew faster (11.3 per cent) than the metropolitan districts (8.2 per cent). Unfortunately we do not know just what this more rapid increase of small cities signifies. A great many of these lie within the bounds of the metropolitan districts already referred to, and their more rapid growth may be largely a consequence of the suburban movement. The tabulations necessary to establish the facts have not yet been made.[2]

[2] The Scripps Foundation is making such a study, but it could not be completed in time to use the results here.

During this last decade, 1930–40, the farm population remained practically stationary in numbers, hence it declined proportionately. The rural-nonfarm population, however, increased by about 3,400,000 or by 14.2 per cent (see Table 2). This is about twice the rate of increase of the nation as

TABLE 2: *Percentage Increase in the Population, Urban and Rural, United States:* 1920–40.

Area	1930–40	1920–30
Total United States	7.2	16.1
Urban	7.9	27.3
Rural-nonfarm	14.2	17.4
Rural-farm	0.2	– 3.9

a whole and considerably more than that of the metropolitan districts (8.2 per cent). This would seem to indicate a considerable movement of people out of cities or at least into less congested areas. Again we cannot pass judgment on the meaning of this movement until we know where these rural-nonfarm folk increased. Preliminary results of the special tabulation just referred to seem to indicate that a considerable proportion of this nonfarm increase took place within metropolitan districts and must be regarded as part of the suburban movement.

Increasingly frequent reference to "blighted" areas in our large cities might lead one to conclude that there was a rather large net movement of population out of the central cities to the suburbs. This does not seem to be borne out by the facts. The central cities claim not only their own excess of births over deaths but still draw some people to them from smaller areas which have higher birth rates, although their drawing power is now less than that of their suburbs.

It would appear, therefore, that one's judgment as to whether the rather slight decentralization of population from 1930 to 1940 is of much significance will depend largely on his evaluation of this suburban movement. Does the suburban movement in its present form yield the results in improvement in living and in changes in social attitudes which we believe decentralization should yield if it is to contribute to the more fruitful development of human nature?

The writer's own feeling is that until there is a definite movement of people out of the metropolitan districts, or until the suburban districts are rather completely reorganized to make a satisfactory home and community life possible to all their residents, we cannot properly consider the suburban movement as accomplishing the purposes aimed at in decentralization. The movement of a considerable number of people from the central cities to the suburbs as they are now organized or, shall I say in their present unorganized condition, does not seem to me to be accompanied by any very significant change in manner of life. It represents merely an individual flight from less to more tolerable surroundings, and it is certainly symptomatic of a revolt against the constraints of congested living, but it involves no clear break with the established values of city living. At the most, therefore, it is only beginning to prepare a population for the establishment of a new social structure in which the values associated with home and children, community life and friends, and living and co-operating with nature will have a better chance to develop than in our cities. Just because it is an unorganized movement of individuals, it can do little more than prepare for the reception of new values; it is no guarantee of their development. But my task here is not to criticize the general values developed in cities but only those which are closely connected with the growth and movement of population. (See Chapters 4 and 7.)

Unfortunately we cannot at present tell much about how war migration affects the decentralization of population. The data available on war migration as of early 1945 indicate a tremendous increase in the population of about 60 per cent of our metropolitan districts as well as in a number of previously semirural counties. In the latter, war services or war manufactures have been located chiefly because of the need for space or isolation. Clearly there will be a large migration of population from such areas despite conversion or reconversion. It cannot be taken for granted, however, that things will go back to "as was" as regards the movement to the large centers. It appears that much of the industrial construction for war purposes which can readily be shifted to peacetime use has been located in the larger manufacturing centers. This large increase in up-to-the-minute manufacturing facilities in these large centers may stop, or even reverse, temporarily, the movement of population into areas outside the larger metropolitan districts. At the same time it may speed up the movement to the suburbs since most of these new factories are on the fringes of the cities.

Since the writer has already expressed very serious doubts whether the present suburban movement will have any significant effect in encouraging many of the more basic social values which have been lost in our cities, he would like to add here that in his opinion large metropolitan districts could be developed which would relieve congestion, would make long and tedious trips to work unnecessary, would encourage community life and good citizenship, and would make home life and the raising of fair-sized families far more satisfactory than is now the case in our suburbs. He believes, however, it will require a far more carefully planned development of the suburbs to bring this about than has yet been undertaken and on a far more extensive scale. The large cities and their suburban areas will never become true nurseries of a healthy national and community life as long as they develop in their present haphazard fashion.

The relation of the city and its suburbs to the natural

growth of population is a point worthy of much more discussion than it can be given here. The large cities of today and their environs in nearly all of the industrialized West have birth rates too low to maintain their present numbers. Furthermore, the lowest of these birth rates are found in the more comfortable economic and the better educated classes. If we assume that it is desirable to maintain about our present numbers, the rates of reproduction in the cities and their suburbs become matters of very great importance, and if there are significant differences between them, the suburban movement might turn out to be of the highest consequence. Unfortunately, up to the present, there is no clear evidence of any substantial difference between cities and their suburbs in rates of reproduction, hence we must assume that, as yet, suburban living is not exerting much influence on the social attitudes which determine reproduction.

The important differentials in reproduction of which we are certain today are: (a) the relatively high rate in the rural population, and particularly the rural farm population, as compared with the urban population; and (b) within the urban population the much higher rate of the poor and the unschooled as compared with the comfortable and well schooled.

There is some little evidence that the families in areas near cities are larger than those in cities, but, as stated above, the situation is far from satisfactory and such differences as there are may be interpreted in ways which cast doubt on differences in fertility in the cities and their suburbs. Thus, even if suburban people have larger families, we cannot be certain that these larger families are not explained by the movement of many couples with children to the suburbs rather than by a higher birth rate to women living in the suburbs. In other words, the suburbs may select the more prolific part of the population rather than encourage greater prolificacy.

TABLE 3: *Number of Children Ever Born per 1,000 White Women, 45–49, Ever Married, for Metropolitan Districts (Urban and Rural) of Cities over 1,000,000. United States: 1940.*

Area	Chicago	Detroit	Los Angeles	New York-N.E. New Jersey	Phila-delphia
Total	2,508	2,679	2,064	2,558	2,743
In Central Cities	2,473	2,642	1,820	2,535	2,729
Other Urban	2,574	2,818	2,220	2,608	2,575
Rural-nonfarm	2,841	2,611	2,498	2,494	3,232

Table 3 shows the number of children ever born per 1,000 women aged 45–49 in 1940 in our five largest cities and their surrounding areas. The "other urban" women ever married and aged 45–49, that is, those living in cities of the metropolitan districts but outside the main city, had a slightly larger number of children than the women in the central city except in the Philadelphia district. The rural-nonfarm women in the Chicago, Philadelphia, and Los Angeles districts had a still larger number of children, but those in the New York and Detroit districts did not. These differences are not consistent enough to permit us to make any general statement concerning facts, but they do cast doubt on the existence of significant differences in the fertility of city and suburban women. In this connection it should be said that Table 3 gives a considerably exaggerated idea of the fertility of city women, since the women of completed families, those 45–49 years of age in 1940, were having most of their children from about 1910 onwards, but before the birth rate fell to the low level it reached in the later nineteen twenties and the first half of the nineteen thirties. Table 4 shows the replacement rate of certain groups of our population as it was in 1930–40 and thirty years earlier (1,000 will just maintain numbers).

TABLE 4: *Net Reproduction Rates, for the United States, Urban and Rural, 1935 to 1940, 1930 to 1935, and 1905 to 1910.*

Area	1935–40	1930–35	1905–10
United States	978	984	1,336
Urban	726	747	937
Rural-nonfarm	1,150	1,150	1,499
Rural-farm	1,661	1,632	2,022

Source: U. S. Bureau of the Census Special Report Series P-1943, No. 5, November 11, 1943. A rate of 1,000 would just maintain population with birth rates and death rates as they were at the given date, with no migration. A rate of 726, therefore, means that under these conditions this group would decline to 72.6 per cent of its present size in a generation, while a rate of 1,661 means that this group would increase to 166.1 per cent of its present size in a generation.

The statement that large cities and their suburbs have birth rates too low to insure reproduction may seem exaggerated to one who has just read in his local paper that *X* had 500 or 5,000 or 10,000 more births than deaths in 1943. The joker here is that although most large cities still have more births than deaths, they have such a surplus only because they have a very high proportion of young adults. According to the 1940 Census, 26.5 per cent of the urban population is aged 20–34 as compared with only 21.2 per cent of the rural-farm population. (See Table 5.) Thus the urban population in the

TABLE 5: *Percentage Distribution by Age, of the Population, Urban and Rural, United States:* 1940.

Age	Total	Urban	Rural-nonfarm	Rural-farm
Total	100.0	100.0	100.0	100.0
Under 5	8.0	6.7	9.3	10.0
5–19	26.4	23.4	27.6	32.7
20–34	25.0	26.5	25.0	21.2
35–49	20.2	22.2	18.7	16.7
50–64	13.5	14.3	12.3	12.8
65 & over	6.9	6.9	7.1	6.6

most important reproductive ages is one-fourth larger than the farm population. This large difference is due to the heavy migration into the cities both from abroad and from the rural areas. Since a high proportion of young adults in any population has the effect of maintaining a relatively high birth rate and a relatively low death rate, the cities appear to have a natural increase long after the number of children born per mother is insufficient to replace the current population. At present it can be safely said that there is no city in the United States having over 100,000 population that would maintain its numbers for two generations if it were deprived of migrants, and after that time most of them would decline by 25 per cent or more in each generation. Many of our small cities are in the same situation but to a lesser degree, for it is generally true today that as the size of the community decreases the birth rate rises. But in spite of the somewhat higher birth rate in the small cities, the only part of our population which is unquestionably replacing itself is the rural population, and even in the rural population the nonfarm group, being rather highly urbanized in many parts of the country, has a comparatively low rate of increase. (See Table 4.)

I would not give the impression that it is a new thing for cities to fail to reproduce. In fact it is usual rather than unusual for cities to depend on migrants for the maintenance and increase of their numbers. Throughout most of man's history, however, the natural *decrease* of cities came about by reason of their enormous death rates rather than by reason of their low birth rates as is the case today. It was not until quite recently, with the coming of the Industrial Revolution, that man learned enough about the control of disease to insure a natural growth of city populations. Prior to about 1750–1800 the sanitation of all cities was so horrible that death rates of fifty or more per 1,000 (they are now ten to eleven in the United States) were usual. A careful study of the registration of deaths in London between 1700 and 1750 led Miss Buer to conclude that the excess of deaths over births amounted to about 10,000 annually. But once even a small measure of sanitation became possible in the cities, the death rate was lowered so rapidly that there soon came to be a substantial excess of births over deaths, and this excess grew as better economic conditions and improved medical care came to supplement better sanitation. As a consequence there was a period in the West from about 1750 or 1800 to 1900, varying considerably both in time and in duration in different countries and cities, when cities had a true and a fairly large excess of births over deaths. Their death rates came under control much sooner and faster than their birth rates. However, this period was relatively short. The birth rate, too, began to come under control in the cities after two or three generations and was soon under such complete control and had been reduced so greatly that most of the large cities no longer had enough births to keep up their numbers on a long-time basis, although their death rates are still declining. This situation, if not already present, is rapidly becoming common to most western lands.

Some facts recently published by our Census Bureau regarding the number of children ever born to different groups of women throw additional light on the failure of our cities to reproduce. In 1910 about 12.8 per cent of the urban white women 40–44 (born 1866–70), who had married and who reported on the number of children they had borne, were childless. In 1940 (women 40–44, born 1896–1900) this proportion had risen to 17.6 per cent or more than one-third; for the rural-nonfarm white women the proportions were 10.1 per cent and 13.7 per cent, respectively; and for rural-farm white women 6.0 per cent and 8.7 per cent, respectively, or just about half as large as among city women. If the women who never married are assumed to be childless, and the same proportion of the married women who did not report on number of children borne as of those who did report are added to the married women reporting no child, we get a total of 27.2 per cent of all urban white women 40–44 in 1940 who were childless, as compared with 24.5 per cent in 1910. A similar calculation yields 20.3 per cent for rural-nonfarm white women and 14.1 per cent for rural-farm white women, as compared with 19.0 per cent and 11.9 per cent respectively in 1910. The increase between 1910 and 1940 in the proportions of the married white women 40–44 reporting who had only one child was as follows: in the urban population, from 13.6 per cent to 21.0 per cent; in the rural-nonfarm, from 11.5 per cent to 17.9 per cent; and in the rural-farm population, from 7.8 per cent to 11.8 per cent.

In most of the large cities the proportions of married women who are childless are even higher than in the urban population as a whole. In 1940 in New York City 18.0 per cent of the white married women 40–44 reporting on number of children ever born were childless, in spite of the fact that a great many of these women were foreign born, among whom childlessness is less than among native born. In Chicago, the proportion childless rose to 18.9 per cent and in Los Angeles to 28.5 per cent. When a proportional number of married women 40–44 not reporting on number of children born and the unmarried women of the same age are added to this childless group, we find that about 29 per cent of the white women 40–44 in New York and Chicago never had a child, and about 36 per cent of those in Los Angeles. Since these percentages of childless women in 1940 relate to women born in 1896–1900 and who were having most of their children during the nineteen twenties, it is safe to assume some further increase both in childlessness and in one-child families among younger women.

It was not a serious matter for the city to depend on the

country for the migrants to keep up its numbers, or even to increase them as long as only 10 to 20 per cent of the people lived there. The 80 to 90 per cent of country people could easily make up this deficit. Today, when many countries have half or more of their people living in cities, the situation is quite different. In 1940 over 56 per cent of the people of the United States lived in urban communities and another 4 or 5 per cent lived in the satellite rural areas of our metropolitan districts and can very properly be regarded as urban in birth rates. Thus about 60 per cent of our total population is now living under distinctly urban conditions and has a deficit of births of about 27 per cent, that is, if it maintained the birth rate and the death rate of 1935–39 and had no in-migration, it would begin to decline by over one-fourth in each generation after its present favorable age make-up had passed. Obviously if the entire nation is to maintain its numbers when our whole population has ceased to have a favorable age make-up and 60 per cent is failing to reproduce by 27 per cent, the remaining 40 per cent must not only maintain itself but add over 40 per cent to its numbers in each generation. In terms of the reproductive performance of an average woman in these two groups, assuming a continuance of 1935–39 birth rates in the urban group (60 per cent), maintenance of our population would require that for each child born to a woman in this group there be 1.92 children born to a woman in the 40 per cent group. Is this a desirable situation even if it would work?

This is not the place to enter into the question of the desirability of a growing, a stationary, or a declining population in the nation. One's judgment on this point will depend on many things which have little to do with the effect of city life on population growth. I do wish, however, to call attention to the fact that in all human history, community life has been organized to insure reproduction to all its members with only minor exceptions (vestal virgins, a celibate priesthood, sisters of mercy, eunuchs of the royal household, and a few other small groups). There have always been some sterile marriages, but they were few and carried a stigma. Furthermore, until quite recently in the West, social sanction was commonly given to a variety of arrangements to insure the perpetuation of the family where the marriage was childless (concubinage, etc.).

Now we come rather suddenly to a form of society in which a large proportion of the women have no children, about 30 per cent in our urban white population, or only one child, about 20 per cent in the same group. This indicates the growth of a new scale of values in our urban community, since the evidence shows that a considerable proportion of all childlessness and of one-child families is voluntary. If this reproductive failure is not voluntary, it indicates a physiological degeneration which is just as deadly. In either event, for any society to have its dominant classes, and no one will dispute the dominance of the well-to-do urban classes in our modern society, lose interest in reproduction, and as a consequence in the future of the community because they have little or no biological stake in it, is an extremely serious matter. If the leaders of any society rather suddenly cease to organize their lives around children and instead organize them around personal desires and the use of goods which they can buy with what they save by not having children, that society will have to face a situation in which it would seem to stand but little chance of surviving. Its people will cease to belong to "the meek who shall inherit the earth."

We know too little to say with any certainty what the social effects on our civilization will be if 30 per cent of our city people continue to have no children and another 20 per cent continue to have only one child. The demographic problem we have already noted. We do know, however, that it is the women with no children or with one child who, in general, set the mode-for-living considered most desirable by the urban community. They are creating a pattern of living which certainly cannot be considered good if we are thinking farther ahead than our own immediate personal interests. No pattern of living which does not include reproduction can be other than ephemeral. A preoccupation with personal ambitions, luxury, and ease of living so great that children have no place in life does not augur well for the future of our civilization. Obviously there is no future for any class, group, or even a nation, if any considerable part of its people persistently refuses to reproduce. There are, in the judgment of the writer, times when the population of a nation should be reduced in numbers in order to make decent living possible; but it is hard to believe that this is now the case in a country so liberally endowed with natural resources as the United States.

I am not a believer in the goodness of large or increasing numbers as such. But in a world which is getting crowded and where other nations because of their large and growing numbers are claiming, and are ready to fight for, larger "places in the sun," I am disturbed when such a large proportion of the people who have been treated most kindly by our society do not find anything in its ideals or in its traditions which is so worth perpetuating that they will rear children to carry on. If the culture and refinement and comfort of modern cities mean so much, or should one say, so little, to a large proportion of city dwellers that two or three children are not worth while, when they interfere with the attainment of these goods, then I am disposed to question the values developed by city living rather than the belief that life itself is worth carrying on. I am disposed to wonder whether the city has not blinded us to the meaning of life rather than to deny that life has a meaning. Certainly no nation or civilization can hope to carry on without children and it seems almost equally obvious that the meaning of life can never become real and vital, except when we are planning for future generations as well as for our own comfort.

4. *What We Are and What We May Become*

PAUL L. VOGT

WHAT we are is largely the result of conditions that have had to be met in a growing country. The direction we have taken, the location of our industries, and the distribution of our population have been determined in part by the slow processes of history, in part by the decisions of persons or groups with reference to immediate chances of profit. Some of the results have been unfortunate, as in the overcentralization of industry in limited areas. Other results are but stages in a movement towards a better balanced economic life.

What we may be depends upon the fullest possible understanding of what we should attempt to achieve in this country. For this purpose, we should know as much as we can about our present condition, some of its causes, and the trends that now exist or may be created for the greater well-being of all. Some of the ideals of economic and social life might possibly be realized through the free play of forces inherent in human association. But most of us act on the principle that, by planning our community life, we can realize our ends much more rapidly.

It may be assumed that conditions produced by the war will not all be permanent. Some of them will. But the tendency will be for the basic influences of the prewar period to become again the prime factors in population distribution and industrial location.

Since where the jobs are determines where we live, we are primarily concerned in this chapter with problems of location of industries. Of the industries, special attention is given to manufacturing because, to a large extent, transportation, commerce, services, and other economic activities depend for their growth upon the processing of raw materials as well as on their production. Production of raw materials is one of the steps in the creation of wealth. Turning of raw materials, through manufacture, into usable forms is another. Where these two activities are found, there is likely to be a large development of trade, transportation, and other services. These provide increasing opportunities for employment and a large wealth-producing, as well as consuming, population.

If we had a map of the United States before us showing where the natural resources are and another showing where the people live we would see a marked lack of balance in the distribution of resources and of people. A little group of states, including Massachusetts, Connecticut, and Rhode Island, the three Middle States (New York, New Jersey, and Pennsylvania), and Maryland, Ohio, Michigan, and Illinois comprise but 9 per cent of the total area. Yet in 1940, 43 per cent of the population, 65 per cent of the manufacturing, and 56 per cent of the total national income was found in these states.

Nine other less well-developed states, including Minnesota, Wisconsin, Iowa, Missouri, Indiana, Florida, Louisiana, and the northern three New England States, with 15 per cent of the area, had 16 per cent of the population and only 14 per cent of the manufacturing.

The remaining states in the West and South, representing 76 per cent of the area of the country, had but 41 per cent of the population, and did only 21 per cent of the manufacturing, yet they had 64 per cent of the mineral resources of the country and 52 per cent of the total farm income.

This regional lack of balance does not tell the whole

story. Seventy-seven million people (56 per cent of the population) live in industrial zones covering only 8 per cent of the total area of the country. And forty-six million are crowded into one hundred and thirty-six counties.

Some of these industrial areas are to be found in the West, as around Los Angeles and Seattle, but most of them are east of the Mississippi and north of the Ohio.

Moreover, in 1939 the areas with the greatest natural resources had the lowest per capita income. Further, the population was densest where the per capita income was high. The war has brought some changes to the advantage of the underpopulated states. Whether this change can be continued after the war is an unsettled question.

This present lack of balance is due to the fact that some older regions, like the East, have advanced into a better balanced economic life. It is also due to the tendency toward congestion of industries and people in already overgrown cities that has characterized modern industrial growth. Other regions, like the South and the West, are still undeveloped.

Economic balance does not call for a dense population in all sections of the country. The mountainous West, the Ozark regions of the South Central United States, the vast arid stretches of the Southwest, and the cold, semiarid sections of the West North Central States do not give promise of a dense industrial population in the near future. There are limitations in agricultural and other natural resources, climate, and water supply that must be overcome before some areas can hope to have a large industrial development.

However, every community that has local agricultural, mineral, or fishing resources has possibilities of turning these resources into manufactured products. Each section and each community must study its own resources and try to develop a natural resource-industrial production economy that is adapted to those resources. Many sections and communities could considerably enlarge their present industrial programs. There are some areas within these regions that promise fine possibilities for development: those for example, in which irrigation or drainage has been or may be developed. Since more than three-fourths of all manufacturing is dependent primarily upon agriculture for its raw material, these areas offer the opportunity to develop a well-balanced agricultural-industrial economy with a fairly dense population.

Census records show that manufacturing, agriculture, forestry and fishing, wholesale and retail trade, and personal and professional service account for 34,000,000, or 75 per cent, of the employed population. Of these, manufacturing with 10,500,000 employed, ranks first in the number of employed workers, and is followed by agriculture, forestry and fishing, and trade, with 8,500,000. Agriculture, forestry, and fishing are decentralized according to sources of raw material. Personal and professional services are related in their distribution to the location of population, although there are likely to be more of these services per capita in centers of population than in rural districts. Of the wholesale and retail groups, retailers, including 6,312,000 of a total of 7,538,000 wholesale and retail workers, are mainly distributed according to location of consumers. Only the 1,106,000 wholesale workers tend to centralize. The records show, however, that only 33 per cent of the total wholesale business is centralized in twenty-seven cities, and these centers include all the leading seaports. Some manufacturing enterprises are doubtless attracted to wholesale centers. But the fact that 66 per cent of the wholesale business is distributed from smaller centers indicates that the factors influencing the two types of enterprise are not the same. The wholesaler locates his business with the objective of economical service to his retail customers rather than with reference to the sources of his supplies.

Manufacturing has a leading place in relation to population distribution, although throughout American history it has been generally accepted that upon the prosperity of agriculture depends the prosperity of the other factors in the industrial system. Changes in the position of agriculture have lessened its relative importance. Agriculture, forestry, and fishing have now taken second place to manufacturing in the number of persons employed. The gross farm income for 1939 was $10,525,000,000 as compared with $24,682,000,000 value added to products by manufacture during that year. Were the manufacturing plants located in agricultural areas, the wealth produced should have been more than doubled.

Of all the industries mentioned, manufacturing, not only because of the large numbers of the population dependent upon it, but also because much of the industrial system is concerned with the handling of manufactured products, may be considered basic to any program of scientific population distribution.[1]

Not all manufacturing is of equal importance in determining population distribution. It has been pointed out by students of the principles of industrial location that the placement of some manufacturing industries is dependent upon the sources of raw material.[2] Others grow up close to the market. Still others are related to more basic manufactures as feeders of semiprocessed material or finishers of products partly processed in other factories, or as users of waste products from basic plants.

The basic manufacturing plants that tend to bring together people in addition to those directly connected with their operation establish the conditions favorable for the

[1] *U. S. Statistical Abstract*, 1941, pages 706, 846.

[2] See studies of principles of industrial location, National Resources Planning Board, *Industrial Location and National Resources*, December, 1942.

growth of wholesale and retail trade, domestic and personal service, and professional service. They also ultimately favor those other forms of manufacture that depend for their success on being close to the consuming public.

Contrary to the impression prevailing in some quarters, the 184,230 manufacturing plants of the United States are already widely distributed throughout the country. This distribution is indicated by the number of manufacturing establishments to be found in each state. The lack of balance between production of raw materials and processing shown on page 87 indicates that this distribution should be much wider than it now is. Present distribution shows general interest, adaptability of industry to location, and ability to survive. The fact that a few industries have given prominence to a limited number of large cities has led to the impression that there is a high degree of centralization of all manufacturing. For example, the motorcar industry recalls Detroit, Michigan; rubber, Akron, Ohio; steel, Pittsburgh, Pennsylvania. A few large industries are highly centralized. But by far the greater number of manufacturing plants which are characteristic of manufacturing in its present stage of development in the United States are widely distributed. The distribution of plants, however, is not accompanied by a similar distribution of manufacturing. The beginnings have been made that might lead to a more balanced ratio between production of raw materials and manufacturing throughout the country. Existing centralization of a few large industries does not indicate at all that existing small plants in newer sections of the country cannot be enlarged or new ones started with good hope of survival.

According to the manufacturing census records durable goods manufacturing plants are more widely distributed than are those producing nondurable goods—68.3 per cent and 34.6 per cent, respectively, in thirty or more states. Of the durable type, fifteen subgroups of iron, steel, and their products, or 55.5 per cent, were to be found in twenty or more states, and 64.7 per cent of machinery subgroup manufactures were reported in twenty or more states. On the other hand, only 49.9 per cent of the nondurable goods manufacturing subgroups were in twenty states and only 24.9 per cent in thirty states. However, one subtype of tobacco is produced in thirty to thirty-four states. Of textiles, one out of thirty-eight subtypes is manufactured in thirty-five to thirty-nine states. And of three subtypes of rubber products, one is produced in thirty to thirty-four states.

The extent of distribution of these industries indicates that it is possible for the great majority of industrial subgroups, both durable and nondurable, to become established in almost any state in the Union. While at present the area included in the New England, Middle Atlantic, and East North Central states is most highly industrialized, the facts as to widespread distribution of some of nearly all industrial subgroups noted above show that there is a possibility of quite general distribution, even of industries that are now highly centralized. It is significant that iron and steel, machinery, food, textiles, and chemicals are all widely distributed.

In 1939 only 2,305 establishments in the United States employed 500 or more wage earners per establishment. At the same time, 168,814 establishments, or 91.6 per cent of the establishments, employed not more than 100 wage earners each. Only 15,416, or 8.6 per cent, employed over 100 persons per establishment. In view of the present development of industrialization in the United States, it would appear that, if size is a criterion, the United States is still, industrially, in the embryonic stage. By extension, the undeveloped areas have a very promising future.

The medium-sized and small plants, which characterize

the less well-developed areas, seem to be able to hold their own during depression as well as in prosperous periods. According to the Census records, the medium-sized plants appear to have closed in a smaller number of cases than either the large or small establishments. From 1929 to 1933, 32.8 per cent of these small plants, 26.4 per cent of the medium-sized ones, and 34.2 per cent of the large ones, closed down. On the other hand, during the period of recovery, from 1933 to 1939, the three size groups showed a quite uniform rate of increase in the number of plants. The increases for small, medium-sized, and large establishments were 30.1 per cent, 27.3 per cent, and 29.4 per cent respectively. As compared with 1929 there were 12.2 per cent fewer small plants, 12.2 per cent fewer medium-sized, and 14.8 per cent fewer large establishments in 1939.

The agricultural and small-town populations have a twofold interest in the distribution of industry, namely as producers of the base materials from which manufactures are made, and as purchasers, often from a distance, of those manufactures.

In the preceding discussion it has been shown that agriculture in the West and South accounts for over half of the total farm income of the United States. No less than 59.9 per cent of the wage earners engaged in manufacture in 1939 were primarily dependent upon vegetable and animal products as a source of raw material, and 76.1 per cent of all manufacturing establishments were dependent upon these sources. Undoubtedly the future prospects of both agriculture and manufacturing in the postwar period are very intimately dependent upon each other. Why not locate manufactures close to their source of supply?

In studying problems of distribution of industry it is important to know something of the present distribution of population. This is of particular significance in relation to manufacturing establishments that tend to locate close to a consumers' market for their products. In 1940, over fifty-seven million of our people lived in communities of less than 2,500 population. Nearly twelve million more lived in places of 2,500 to 10,000, seventy-nine million lived in rural communities or in places of less than 25,000, and of the fifty-three million living in or around central cities only forty-two million lived in the central places. The consumers, when they can, live in the less congested areas. Were manufactures located in smaller places, doubtless a large number of those now living in large cities because their jobs are there would be glad to live under more favorable conditions elsewhere.

Since only about 40 per cent of the people in the United States live in central cities of above 25,000 inhabitants, the need of the manufacturer who desires to be near the consuming public for locating in a large city seems to be decreasing. Improved transportation facilities also tend to decrease the need for urban congestion of industry. Nearly two-thirds of the consumers live in suburban, small-town or rural areas. Moreover, with the exception of a few establishments that depend for effectiveness on vast production lines or great expenditures in machinery and equipment before production can begin, a very large proportion of United States manufacturing plants are relatively small and are quite generally distributed throughout the United States. The embryos of considerable industrial development have been established on a wide scale and in great variety over the country. Encouragement of plants already in existence may do as much to stimulate industrial growth as the establishment of new industries. From the point of view of distribution of small manufactures, the outlook for attainment of better balance between agriculture and industry is hopeful.

The problem of decentralization is not principally one of reducing the city to improve the general public social and economic health. It is this in part, but a large part of the problem consists in helping to develop small-town industries. An important, if not the most significant, objective is to discover the direction toward a well-balanced rural-industrial development. The breakup in the industry of large urban centers should be considered only to the extent that excessive congestion of industries results in national economic and social ineffectiveness and loss.

The number and distribution of manufacturing plants of practically all types over a considerable part of the United States indicates that these plants have had, in the judgment of persons interested in manufacturing, a chance to survive. If this were not so, the risk would not have been taken. No data have been found bearing on the rate of turnover of manufacturing plants by states. The records show that, with the exception of the depression years, the number of manufacturing plants of various sizes has remained fairly constant. Since the depression and up to 1939 the number of smaller establishments was again on the increase.[3]

The need for streamlined assembly plants in some industries, heavy preliminary plant equipment, and production for both national and world markets appears to justify the establishment of some very large centralized plants. The fact that a large proportion of manufacturing establishments are small and widely distributed indicates that large size is not a requisite of success. The indications are that the medium-sized plant is, for most industries, the most profitable.[4]

Manufacturing in the United States is still in its infancy.

[3] For purposes of this study a medium-sized plant employs 101 to 500 wage earners. No official definition of what is a large or small business has yet been established. Even employment of 500 is sometimes used as a dividing point between large and small business. One hundred is also commonly used. For further discussion see U. S. Department of Commerce, *Survey of Current Business*, May, 1944.

[4] Reports of the Federal Trade Commission.

The record of the twenty years preceding 1940, it is true, does not show any marked increase in the number of plants, persons employed, or value of products manufactured. But the increase in population that may be expected for several decades to come, and the increase in national income that may result from the war should create a considerably increased demand for manufactured products. The industrial development of other nations should encourage international exchange of manufactured products. These are among the reasons for believing that manufacturing should expand, at least in quantity of goods produced, during the coming years. The large numbers of widely distributed small plants present every appearance of being new ventures in an expanding economy rather than the vestiges of a passing, once-widespread manufacturing activity. These smaller enterprises should make an ever-increasing contribution to the well-being of our people.

The possibilities of fitting industries primarily dependent on mineral products for raw material into a balanced agricultural-industrial relationship are more limited for some groups. Petroleum and stone manufactures tend to be located at the source of the raw material. The smelting of ore is also likely to be located at the source or at places where there is favorable combination of raw material and power. Later processes appear to be relatively free of limitations of source of materials and to be more subject to market conditions or other factors. The wide distribution of manufacture of mineral products shows that, with the exception of first steps in the fabricating process, there is little to prevent freedom of movement of industry or its location according to principles of balance between agriculture and manufacturing.

The records show that, even in industries dependent upon agriculture, there is not a necessary relation between location and source of supply. Data available from the National Resources Committee[5] throw some light on the problem of manufacturing distribution. Of two hundred and eighty industries studied by the Committee, the location of only thirty-three was determined by the source of raw material. Of these, three were classed as large industries, three as medium, and twenty-seven as small. Sixteen of the industries were primarily related to markets for their products. Two hundred and thirty-one industries, or 82.5 per cent, were located for reasons other than proximity to raw materials or markets.

These data, if typical (and they were studied by the National Resources Committee because they were assumed to be so), are significant in that they indicate an apparent freedom of manufacturing generally from raw materials or market considerations in seeking location. Industries that depend largely on markets will continue to seek locations where population is dense or may be easily and economically reached. In the study of possibilities of developing manufacturing industries, the discovery and utilization of local raw materials deserves major attention. The possibility of developing industries relatively free from either source of material or market influence is also significant.

What has delayed industrial development where it is still of less importance than the production of raw material? A number of factors are responsible. One of these is that, in the development of new areas of the country, the production of raw materials available appears to be economically the most profitable avenue of investment of funds. The basic reason for the failure to develop manufactures in newer sections of the country lies in the relative returns for capital. Past experience appears to indicate that there is a fairly uniform sequence in the development of the various phases of

[5] National Resources Committee, *Structure of the American Economy* (1939), page 265.

economic production. When a new country is opened up, the first step is inevitably the development of the natural stores of raw materials. Whatever capital may be available can be most profitably used in this development. Only a very limited volume of manufacturing is undertaken. Since population is likely to be relatively sparse, manufactures that depend on location near to consumers would have little hope of survival. Those establishments that perform the first steps in processing, such as smelting of ore, are most suitable for development. This has been the history of all sections of America. In many sections the first, or raw-materials-production stage, is still predominant or at least very important.

After whatever capital is available has been used in the opening of mines, the clearing and drainage of land for agricultural purposes, the building of homes for farmers, miners, commercial, and other groups, and the starting of trading centers for the accommodation of settlers, a surplus of local capital begins to accumulate. The opportunities for investment in the production of raw materials decline. Then the beginnings of the industrial stage of development occur.

It appears that industrial development tends to be delayed beyond the point or time when it promises returns as good as those from capital invested in the production of raw material. In the first place, as compared with conditions in new areas, the older, more advanced sections of the country have developed manufacturing establishments beyond the small initial stages that are often costly. Managers have had experience relative to size of plant most profitable to operate and have adjusted themselves to principles of economy in production. They have developed well-known national brands of products. They have had experience in marketing and finding sources of raw material at most advantageous prices. They have developed a labor force with the requisite skills and with attachment to the locality. Operating capital from financial institutions is more readily available to going concerns located near those institutions. The new sections in many cases lack not only a labor force and markets, but leadership fitted by training and industrial development. Habit leads the native population to seek opportunity of making a living in existing, prevailing types of economic activity, and these have been concerned with the production of raw materials.

But these seeming advantages of the older communities are more than counterbalanced by others favoring the newer sections. Local investment capital is increasing. War experience has shown that a trained labor force can be created in a reasonably short time. The unfair practices that once handicapped small businesses are known and are being controlled. There are economies in being located close to sources of raw material. New methods often favor the new business. Trained business leadership with aggressiveness and imagination can find an opening in the small, decentralized manufacturing establishment. Large distributors of consumers' goods are in some cases inclined to favor manufacturers in decentralized areas, since widespread purchasing power is thus increased. The outlook, in short, is good for the decentralized business.

Discriminatory transportation charges have long been held responsible for delay in the industrial development of the southern and western sections of our country. In the days of intense interregional rivalry for development, railroad rates undoubtedly had great influence in determining the location of industrial centers. The advantages of trunk-line shippers in railroad rates may be illustrated by some examples of rates between practically equidistant points from southern territory to points in trunk-line territory and the reverse; and between typically equidistant points within each of these areas.[6] From Knoxville, Tennessee, to Columbus, Ohio, 393 miles, the first-class rate was 141 cents for 100 pounds. From Baltimore, Maryland, to Warren, Ohio, 393 miles, the rate was 94 cents, an advantage of 47 cents a hundred. From Charlotte, North Carolina to Saint Louis, Missouri, 801 miles, the rate was 198; from New York City to South Bend, Indiana, 803 miles, it was 142, or an advantage to the trunk-line shipper of 56 cents a hundred. A comparison of rates for goods shipped into southern territory from trunk-line territory shows that from Albany, Georgia, to Waycross, Georgia, 112 miles, the rate was 82 cents, while from Enfield, Illinois, to Nortonville, Kentucky, 104 miles, the rate was 64, an advantage of 18 cents; from Barnesville, Georgia, to Guthrie, Kentucky, 398 miles, the rate was 142. From Chicago, Illinois, to Guthrie, 391 miles, the rate was 116, an advantage of 26 cents a hundred.

The disadvantages were even greater for other sections. From Fort Worth, Texas, to Cincinnati, Ohio, 962 miles, the rate is 244 cents a hundred; from Cincinnati, Ohio, to Portland, Maine, 964 miles, 154 cents, an advantage of 90 cents. From El Paso, Texas, to Springfield, Illinois, 1,236 miles, the rate was 330; from Springfield to Lewiston, Maine, 1,216 miles, 173 cents, an advantage of 157 cents a hundred. From Wichita, Kansas, to Youngstown, Ohio, 914 miles, the rate was 220 cents; from Dayton, Ohio, to Portland, Maine, 911 miles, the rate was 149 cents, an advantage of 71 cents. From Hutchinson, Kansas, to Chicago, Illinois, 649 miles, the rate was 184 cents, and from Chicago, Illinois, to Williamsport, Pennsylvania, 653 miles, the rate was 127 cents, an advantage of 57 cents. From Sheridan, Wyoming, to Peoria, Illinois, 1,115 miles, the rate was 289 cents; from Springfield, Illinois, to Boston, Massachusetts, 1,109 miles, it was 168 cents, an

[6] *Interterritorial Freight Rate Problem of United States*, H. D. 264, 75th Congress, 1st Session.

advantage to the trunk-line shipper of 121 cents. Commodity rates show similar advantages to trunk-line or official territory.

The breaks in rates at territorial border lines have given distributors just within official territory an advantage over distributors near these borders on the other side. Recent changes made by the Interstate Commerce Commission have ameliorated this condition so far as distributors are concerned but have accentuated manufacturers' advantages in official territory.

The influence of rates on the development of the different sections of the country is summarized in the report on the interterritorial freight rate problem mentioned earlier. "The conditions which have been pointed out . . . have been slowly developing over a long period. During all this time the interterritorial rate problem has gradually been crystallizing into its present form. For nearly three generations, or since the earliest days of the railroad, the country has been accepting influences and conditions which have made a workshop region of one section of the country, with the other sections, as marked off by the present territorial rate boundaries, being largely in a position of contributors of raw materials and semifinished products to the industrialized region. This has been true, despite the fact that certain areas of the present rate territories possess in one way or another excellent opportunities for the development of industries which would bring to those areas a greater measure of economic and social stability. Otherwise, if we continue a heavy concentration of manufacturing in one region with its ever-present concomitant of drainage of raw materials from the others, we may expect eventually to have an exhaustion of natural resources in many areas without leaving to the population which will have devoted itself to exploiting such resources any great tangible means to represent the wealth which they will have shipped away. There is too much evidence on every hand of this very thing at the present time."[7]

The history of attempts to start new industries in undeveloped areas is full of experiences that today tend to act as deterrents to the development of sound enterprises. The high-pressure salesman who has sold stock in a proposed factory has been too well known in many communities all over the country. The fly-by-night enterpriser who collects large funds from a local civic improvement association and then moves his factory to another village is remembered by many investors. Inexperience, too, in the fundamentals of manufacturing enterprise and the ease with which a new concern can be milked of its resources have been costly to the well-meaning but unwary in the West and South. If these regions do not respond as rapidly as others to the challenge of industrial development, their failure can be charged in part to sad but misleading lessons learned in the past.

Moreover, it has been said that in some instances manufacturers who have already established themselves in local communities oppose further industrial expansion because of their fear of losing control of the labor market. Hardly less important is the apprehension, sometimes expressed, that the development of new business may be handicapped by the control of investment funds by large banking organizations whose officials are also interested in highly integrated, large-scale enterprises. Lower interest rates for large concerns, the centralization of control of patents, and unfair business practices are also subjects of complaint.

Although the location of many large industrial centers was once determined by the ease with which steam power could be generated, the transmission of electricity has now freed many industries to locate themselves at considerable

[7] *Interterritorial Freight Rate Problem of United States*, H. D. 264, 75th Congress, 1st Session, 1937.

distance from the source of power. The full implications of this new freedom have yet to be felt and appreciated by the underdeveloped regions of the country. In the not too distant future, when the expansion of American industry into new areas is an accomplished fact, electric power will have played the role of pioneer.

Thus far the factors noted as having influence on manufacturing distribution have all been of economic or political significance. Of greatest importance, however, is the quality of local leadership. As has been pointed out, with the exception of some industries that must be located near the source of production of raw materials or at ports of entry of such materials, most manufacturing has been freed from other influences that previously determined location. The resourcefulness, vision, energy, and training of potential enterprisers count much more largely now than ever before.

In this chapter we have pictured the resources of the country and presented evidence that much of the West and South is still young so far as industrial development is concerned. Natural resources are known and the capital goods necessary for their production have in many cases been installed.

Some parts of the West and South are admittedly handicapped by reason of ruggedness of the land, or by lack of water for industrial use. But the combination of agricultural, mineral, and water resources in a considerable part of the area justifies the belief that a great agricultural-industrial culture is destined to cover it.

What we may become may be predicted from what now is in states like Ohio, Pennsylvania, New York, and southern New England. These areas are not overdeveloped. They may be badly developed in so far as the earlier growth tended to centralize in and congest large urban centers.

The maps showing distribution of typical resources indicate that the South has the basis for a fairly evenly distributed population in small and medium-sized industrial communities. The agricultural resources do not vary greatly from district to district. In the West, because of the large areas suitable only for raising of livestock or for timber production, a different picture is presented. Where irrigation or drainage is possible, giving the basis for a varied production of raw material, an industrial superstructure may be built. Similar industrial growth is possible where mineral products or fisheries are centralized. In time we may expect to find many areas of considerable size intensively developed on a scale similar to that now found east of the Mississippi River and north of the Ohio River. Much of the area in the agricultural states east of the Rocky Mountains can support a much greater industrial population than it now has.

The underdeveloped areas offer large opportunities for industrial leaders of vision who have faith in them. They have need for a greatly increased population which not only will produce raw materials but which will fabricate these materials into forms usable by a large population of producers of raw materials and service groups and their families. The people of these areas need some encouragement and a fair field for their enterprise. The newer sections in time should be dotted with many small and medium-sized centers of industry in and around which will be found the highest type of American life.

5. *Biological Truths and Public Health*

JONATHAN FORMAN

MEDICAL SCIENCE, with the assistance of sanitation and public health, has worked a vast improvement in the lives of men during the past century. In general, it can be said that humans today are less beset by disease, are better nourished, and are living to riper years than their forebears could hope for in the period when the industrial age was yet young. Within certain limits, medical science has made it possible for the average individual to expect health and long life as normal endowments, and frequent or pestilential illness and a short life span as exceptions. But certain of the accomplishments of science have been offset by the peculiar nature of man's choice of living conditions.

Just as the size of armies was once limited by the lack of sanitary resources, so that of the medieval city was limited by the problems of sanitation. The machine age, which ushered in the mass production of traditional and new conveniences of all kinds, also made possible an abundance of sewers. By means of them cholera and many enteric diseases which had held city growth to a minimum for generations were largely eliminated. With modern plumbing came a clean water supply, and typhoid fever all but disappeared. Thus were conquered the main obstacles to city growth. The sequel, however, is less reassuring.

Today, because of improved health conditions, cities have, in many instances, grown so large that they are no longer healthful places in which to live. The paradox almost contains its own explanation. Ills that have sprung from man's mismanagement of his environment have reversed some of the gains painfully purchased by generations of scientific effort and industrial advancement.

In candid truth, the city is not a healthful place in which to live. Its inhabitants do not live as long or as happily as dwellers elsewhere. More of them go insane. They are sick more often. Finally, and perhaps most disastrously, they fail to reproduce themselves. The city, as a consequence, must import nearly one-half of the next generation in order to maintain its size and satisfy its civic pride. If pure surmise were one of the legitimate functions of science, it might be interesting to contemplate the fate of a metropolis in an age when new blood could no longer be piped to it from the country.

Large-scale industry is a new activity, and people have swarmed around its factories and offices to get jobs. As a result, cities have grown rapidly, without intelligent planning; indeed, the rapid and disorderly growth is responsible for many of the undesirable elements in city life. It has produced poorly designed streets that are congested with unnecessary traffic and full of danger; an atmosphere that is filled with noxious gases, dust, dirt, and smoke—a blanket of smog which shuts out the health-giving sunshine. Neighborliness has been destroyed by overcrowding. Disturbed mental poise and disordered social habits have been engendered by loss of sleep, malnutrition, and frustration. The loss of sleep is due largely to night life and noise; the malnutrition to bad food habits and ignorance; and frustration must always be the price, no matter what the form of the social order, that men must pay for living in cramped quarters and interdependent environments. Neither an apartment nor a tenement is a suitable place to rear a child or keep a dog.

There is no more accurate measure of the healthfulness of the surroundings of an animal than to determine the effect of its environment upon its fertility. The farmer has recognized this fact in a practical way for a long time. When his cow, ewe, or sow becomes a "shy" breeder, bearing young only every second year, he knows that something has interfered with the animal's nutrition so that she must take off a whole year to build up reserve enough to reproduce. So, too, the horticulturist knows, when his orchards begin to bear fruit only every other year, that his soil has become depleted. The gardener recognizes the same fact when he says that he must let his land "rest." Both the gardener and the horticulturist understand that it takes an extra year to weather out the minerals in the soil and make them available to the crops. Thus it is clear that any social order is sick when the total number of children who grow to adult life is not equal to the number of people in the preceding generation.

With this thought in mind it is significant that in 1940 the reproduction rate of the rural-farm population as a whole was twice as high as that of the urban population. This means that the cities lacked 26 per cent of maintaining their populations without immigration, whereas the rural areas had an excess of 59 per cent over what was necessary for the permanent replacement of their populations. It is apparent that the failure of city folk to reproduce is the result of a good many factors, but it is believed that malnutrition in the woman of child-bearing age may well be the most important single factor.

In addition, it must not be overlooked that marriage fails in the cities in an alarming number of instances, and the high divorce rate is no small item in the failure of cities to reproduce their populations. Family life attempted in cramped quarters, "with never a room to call one's own," calls for a capacity for social adjustment, mental poise, and tolerance which few people possess. Consequently there are about twice as many divorces in cities as among rural people. The unfortunate ending of many marriages in the city is the result of many forces—lack of home life, irritability, financial strain, ambition, malnutrition, and neuroses.

Fortunately, in recent years the infant-mortality rate has been much lower in metropolitan areas as a whole than in rural areas. This difference is due largely to better care of babies in cities; and the use of refrigeration has contributed much to the record. Not all city babies, however, have the same chance of survival, for babies of the poor are more likely to die in infancy than those of the well-to-do. Poverty, of course, goes along with malnutrition, depleted biological reserve of parents, poor food habits, poor food, and poor facilities for handling food—all of which are hazardous to babies.

Once infancy is passed, there is little difference between city and country so far as the dangers of childhood are concerned, for in this group the death rate for children is approximately the same. After the individual attains his majority, it is much safer for him to live in the country. This is attested by the fact that both the death rate and the sickness rate for adults are definitely lower in the country. (In Ohio, for some reason, after one has reached the age of eighty, it is safer to live in the big city—at least the death rate for the very old people is lower.)

Life expectancy of the nation as a whole is a little over sixty-four years, and although the average is much ahead of that of any other nation, it falls short of what we have a right to anticipate. It is generally accepted that the life span of any animal should be approximately five times the length of the period required to reach maturity. There is considerable evidence to support the conclusion that man should live a healthy, active life until he is a hundred. While this is an interesting theory and should serve to direct more attention to health and physical fitness, the practical fact strikes us in the face that in the city in normal times no one wants to hire a man who is much older than forty-five years. Confusion, competition, disappointment, and unhealthful factors in surroundings do make many city residents seem old at fifty; and the number of persons in the United States who are over sixty-five years is growing year by year.

Thus comes into existence a social problem of increasing magnitude—the care of the aged. Chronic illness and the necessity for nursing care are the most important problems here. They occur more frequently among persons over fifty-five years of age in the city than among those of corresponding age in the country. Old men and women in urban districts are a care to younger members of their families and often to other people as well. They are usually in the way. The city has little to offer them, and indeed they have little to offer it. The increasing number of "rest homes" and "homes for the aged" testifies to the plight of these poor souls who must live in institutions. On the farm and in the village, on the other hand, old persons have their chores, gardens, flowers, and other responsibilities commensurate with their capacity for work. In the country, grandfather and grandmother are a part of the economic pattern.

If the city has come off second best in the discussion thus far, it should be pointed out that we are concerned here with health and medical realities. The country is not Utopia, nor can it be made so except by the exercise of man's ingenuity. Nevertheless, the very naturalness of most rural and small-town situations precludes stresses and strains and causes of disease which are all too inevitable in the metropolis.

The stresses and strains of city living are of many types. Those that develop from transportation causes are immediately apparent. The disorderly growth of the modern city has come about in no small part from the desire of people to work in one section, live in another, and, very often, shop in a third. This purposeless travel has resulted in the overcrowding of streets originally designed for horse and carriage—even streetcars came after many of our streets were laid out. No medical man can fail to assess the damage to nervous systems, the fatigue, and the reduction in the number of recreation hours that result from the transportation factor alone.

Injuries and fatalities cannot but occur in consequence of the uneconomical congestion of city streets. Aside from the hazards of congestion, malnutrition and jaded nerves of city dwellers are important factors in injuries resulting from traffic. In metropolitan areas, traffic victims are frequently older pedestrians. Accidents occur most frequently in the evening—a fact easily attributable to haste, fatigue, and night blindness on the part of both motorcar driver and pedestrian.

Partly because of transportation difficulties, partly because the facilities are not available, the modern city dweller gets little out-of-door recreation. Too much stress cannot be put on the value of such recreation, for much of man's happiness and well-being should come from hours spent in the open air. The efficiency of work and the moral plane of existence are normally improved when we get close to nature.

If the need of the adult for recreation is great, the case for the youngster is even more pressing. All children need playgrounds and other recreational facilities: the offspring of the wealthy, so that they may enjoy the benefits of competitive sports; those of the poor, so that they may develop healthy bodies and sound attitudes. The fact stands out that healthy children develop good character as an expression of good health.

When recreation in the out-of-doors is not possible for youngsters and adults, they turn to artificial forms of amusement. Far from being a compensation for what is lacking, these forms of amusement serve as detrimental influences upon the individual from a physical standpoint and contribute to bad habits and a general weakening of physical and moral stamina.

Hardly a city in America has enough parks and playgrounds. The further alarming fact is that few cities, if any, can manage, within the limitations imposed by their very physical organization, to remedy the defect. In the face of this circumstance, there is a strong argument for urging the individual to seek, for himself and his family, those conditions which are more nearly satisfactory—a place in or nearer the country where normal values may be attained.

Another ever-increasing menace to the health of our people is noise. It can now be accurately measured and its loudness recorded. In this way, much has already been learned about its harmful effect upon man. Any unexpected sound above the rustling of the leaves on a tree is annoying to the human mind. If this be true—and it is—then it is no wonder that city life is hard on our nervous systems. Urban life—with its crowding, its mechanical devices, its motors and its factories, its people constantly on the go—some coming home, some going to work, and some trying to sleep—by the very noise alone that it produces is detrimental to a great many persons.

We know that noise causes fatigue and impairs the hearing. It produces emotional disturbances, neuroses, and ill health. It prevents sleep and proper rest for all too many persons. While much could be done to lessen the constant strain and drain upon the nervous system of the city dweller, cities will always be noisier places than rural areas.

Noise in industry can be controlled, at least to the point where it does not interfere markedly with the efficiency of the worker. In the first place, most noises in a factory are rhythmic in character, and rhythmic noises are not as hard on the nervous system as those which come at unexpected times. It does not take an unusually loud noise to raise the blood pressure of a surprised listener twenty points in less than twenty seconds. After a sudden noise about equal in its intensity to that produced by the explosion of a small fire-cracker, the smaller arteries of the body contract and so narrow the opening of the vascular bed that it requires a pressure of from 20 to 25 per cent above the regular pressure to get the blood around and through the body.

This, of course, means that upon such an alarm a whole chain of events is set in motion looking towards the defense of the body against an enemy which actually never materializes. Violent reactions of this sort leave the defense mechanism of the body quite depleted.

At work the city dweller is not so adversely affected by the noises of his office, store, or factory. They do, however, interfere quite decidedly with his efficiency and enter in as one more cause of his frustrations and disappointments. Most writers on the subject cite the instance of the factory in which experienced workers were assembling temperature regulators. The center of this activity was a room next to the boiler room. The workers here became an industrial problem, for there were sixty imperfections in approximately every eighty regulators. When these workers were put in a quiet room, they were able to assemble one hundred regulators in the same time that it had taken to assemble eighty, and there were only seven imperfections in the hundred regulators, as compared with sixty in the eighty regulators assembled in the noisy room.

Again, there is the well-known study of the effect of noise upon the efficiency of office workers. In a certain business office when the noise level was reduced 14.5 per cent, the efficiency of the average worker was increased 8.8 per cent. At the same time, the mistakes of the typist were reduced 29 per cent, while those of the machine operators were reduced 52 per cent. Interestingly enough, the labor turnover dropped 47 per cent and absenteeism also fell off 37.5 per cent. Metabolic tests in other studies have shown also that experienced typists use more energy when working in a noisy office than in a quiet one.

In the noisier industries considerable damage is done to the hearing of the workers. Both labor and management are working to reduce and eliminate factory noises as far as possible, as is the Noise Abatement League. Still, in the very noisy industries impairment of hearing is taken for granted. For instance, unless an applicant for a job as operator of a drop forge is partially deaf, he often is not hired. The degree of deafness is assumed to be a measure of the applicant's experience. No deafness, presumably no experience!

All of the noises of the city, rhythmic or not, are much more upsetting to the hearing and the other parts of the nervous system than any sounds encountered in the country, but so accustomed to them is the city dweller and so constantly is he at work shutting them out of his mind that frequently he cannot sleep in a farmhouse because of the quietness. Yet the conditions prevailing in the country are those that were intended for the welfare of man.

Man has often gone without food for days without serious injury to himself, but the lack of sleep for three days is attended with serious symptoms. In fact, a man can live six times as long without food as without sleep. He lives almost as long without water as without sleep. Yet most of the people who live in our cities keep cutting off a few extra minutes of rest each year, going to bed a little later and getting up a little earlier. Although rest is essential for man, millions of our people try to get this vital renewal of their energy while their jagged nerves are being bombarded continuously by all of the noises of city life.

Man came indoors only yesterday. For centuries he developed in the brightness of the out-of-doors. True, he slept in caves, but he did not work there, not even after he discovered fire with which to heat and light his home. Oil lamps were used for lighting through most of recorded time, but they were not worth much until the round wick was developed in the eighteenth century. Candles were for the rich and the churches. Since man has developed in the out-of-doors, his eyes through thousands of years of experience are adapted to its brightness. For untold time he went to bed with the sun and rose with it. As cities have developed in recent years, man and his eyes have become more and more the victims of darkness. Man turns night into day on his schedule and plays and works mostly under artificial light. While there has been great research in illumination, and the illuminating engineers know how to give us the brightness of the out-of-doors in our homes, schoolrooms, and places of work, yet few such places have anything like adequate lighting.

The tall buildings in which we work are poorly provided with light for the most part and are built too close together, with too few windows. Our houses, apartments, and tenements are dark. Our habits and hours of working are bad. The result of all of this eyestrain to which the average city dweller subjects himself is ill health. The eyes are a true outcropping of the brain itself. Misuse of them registers at once in the nervous system. When the eyes are suffering from strain, the resulting disturbances may be referred by the nervous system to any part of the body. Eyestrain is the most common cause of headache and of functional dyspepsia. But more important to the average city person than the ill health it produces is the pessimism, discouragement, and irritability that appear in his fellows because they use their eyes too much and too often after the sun goes down in places that are too dark.

Only too well do we know that the air of our cities is contaminated. Smoke, noxious gases, and vapors are set free to escape into the air in ever-increasing quantities. A whole volume might be written concerning the irritation and poisoning that may result to the city dweller as a result of his constant exposure to the gases and vapors of industry. It will suffice, however, to point out that even in those factories in which these conditions are under the best of control, there is still more hazard to optimum health and longevity than is encountered in the country.

In the traffic-congested streets of the city, the air is loaded with more carbon monoxide than is good for anyone. It is a contributing cause of the city dweller's headaches and raw nerves. Those in the streets are not the only persons exposed, moreover, for, in the winter months at least, much of the air with which the tall buildings are ventilated is taken in at street level and forced through the building. In this way noxious gases may spread to the highest rooms. Not much headway has been made with respect to the use of vertical exhaust pipes for automobiles, although they do have much merit as a remedy for the situation.

Dusts arising from manufacturing processes are likewise harmful. They are produced by mechanical actions such as grinding, crushing, cutting, drilling, and so on. Inorganic dusts infiltrate and discolor the lungs and block the lymphatics of these organs so that absorption is interfered with. They may contain free silica, so that their inhalation may produce silicosis. If this condition becomes extensive throughout the lungs, the resulting disability becomes total and often permanent. It is for this reason that every effort is now being made to keep silica out of the air of industrial plants.

The atmosphere of coal-burning cities contains a great amount of sulphur. For instance, it has been calculated that the rain, snow, and dust of Minneapolis drop four hundred pounds of sulphur dioxide on each acre each year, while at the headwaters of the Mississippi in the same state, the annual drop was estimated at only five pounds per acre. It has also been shown that this atmospheric sulphur is carried by the wind as far as fifteen miles into the country—an important fact to the man selecting a suburban home and garden. With all of this sulphur in the air, it is little wonder that drapes, carpets, and clothing are eaten up by the acid formed by these fumes and the moisture in the air. It must also be a matter of considerable adjustment for the mucous membranes in the respiratory tracts of the average city dweller.

There is dust in the country, of course, and dust storms in some parts of the country are undoubtedly bad for the health of the persons living there; nevertheless, ordinary dust found in rural areas is organic dust, which is easily absorbed and causes no special trouble except to those who have an allergic constitution and have become sensitive to some particular organic ingredient of the dust.

The nose is called upon to remove these dust particles as the air is pulled through it into the lungs in breathing. It filters from twelve to seventy-two quarts of air every minute, depending upon how active the individual may be. It is apparent, therefore, that the filtering mechanism of the nose does a big job remarkably well. Even it, however, often becomes swamped and bogs down under the load put upon it by the smoke-laden, dusty air of the city.

The most unhealthful aspect of city dust is the way in which it combines with smoke and fog to hang a curtain between the city folk and the sun from September until June.

This pall is often so thick that it hides the sun until well along in the day. In this way it blots out many of the health-giving rays that should normally reach us. This blanket of smog, as it has been called, does much more harm to the health of the average city person than he has any idea of.

The building of a house in the temperate zone presents a paradox. It must keep out the cold air in winter and at the same time do something about the constant contamination of the room air by its occupants. In addition to the carbon dioxide which they are continuously breathing into the air, tobacco smoke and other contaminating influences must be removed. The inadequate floor space in the average city home accentuates this contamination.

Air conditioning, as we now have it in summer in office, store, workshop, and movie, is no blessing to the person who must work in these cold, damp places all day and then go home to rooms which have soaked up heat day after day until their stifling air is well-nigh unbearable. Night after night the person tries to make this violent adjustment and night after night he fails. His well-baked bedroom with its piping hot air is bad enough by itself, but now he must stand the shock of the change from the artificial cold to his own hot room. All too frequently the city dweller then loads his stomach with ice-cold drinks and calorie-rich cold food, taking on the heat-producing task of burning up all of these unnecessary calories. Small wonder that he lies in his bed and sweats, half sick and sleepless, until it is nearly time to get up again. And if he was once a country boy, he may remember how cool the old farmhouse was after the sun went down.

Even if we could live and work in a monotonous artificial climate, we would stand a very good chance to lose many of the advantages of living in the temperate zone. Recent investigations have disclosed many facts about the effect upon our health and capacity for work of both weather and climate, and we find that we need the creative value of a stimulating winter alternated with a refreshing summer. Thus careful consideration should be given to the effect of constant air conditioning before we decide to adopt it. At any rate, it will be a long time before air-conditioning equipment will be put into mass production at such a rate that it can be had for the average city or suburban home. And it is certain that never will anything we make ever equal, for health and comfort, the shaded house set in a shady lawn well away from other houses on a farm or suburban acreage.

The temperature of the home must likewise be adjusted so that its inhabitants can live and work there on cooler days. This means that it must be warmed. We have made great progress in heating, but still are far short of the ideal. First came fire in the cave, with its uncontrolled smoke blinding and choking the family. The chimney proved a great invention, for it did much to improve the ventilation. With the chimney came the fireplace—a pleasant but not very effective method of heating a room. Next, the stove brought many additional advantages. And when the central heating plant was introduced, so that all rooms in the home could be heated alike and at the same time, a crushing blow was dealt to colds, pneumonia, and more especially tuberculosis. There are, however, definite disadvantages to central heating, be it hot air, hot water, or steam. In American cities it is the rule to find rooms overheated and the air so dried out by the heating process that the moisture of the skin is whisked away at once. This rapid evaporation makes its victim feel cold no matter what the temperature of the room may be. The comfort of a room is determined by the cooling power of its air, which depends in turn upon four factors—the temperature, the humidity, the velocity of air movement, and our own radiant heat.

A well-known allergist studied the humidity of some apartment houses in the Chicago area and found that in midwinter the air in these rooms was actually drier than that on the Sahara Desert. Students of allergy and other respiratory diseases are certain that overheated, overly dry air is a factor in the development of sinus disease and allergy. The heating problem has not yet been solved, but the man in the country in his own home with a modern furnace and a fireplace, with 150 or more square feet for each member of his family, has the best heating that is available.

Perhaps far more important than the physical disabilities growing out of city life are the mental ailments that may be traced directly or indirectly to our habits and customs as city residents. Mental attitudes are a most important factor in man's health and happiness, and one of the first of mental ills, frustration, is much more common to the city dweller than to the country man—there are so many things that he wants and cannot have. More than half of the patients who appear in the office of the physician in a modern city are not sick with any of the diseases described in a textbook on the practice of medicine. They represent a group of individuals seeking an escape from the realities of life, who yet are not to be rewarded in their search. Every physician is all too well acquainted with this type of patient. His suffering may be real, but his physical examination and his X-ray and laboratory findings are always negative. He is one of the convalescents who do not convalesce. There are recurrences and relapses that no one can explain. These cases do not yield to the ordinary, approved methods of practice. Every physician has many of these patients with persistent headaches, convulsive seizures, fainting fits, heart pains, disabling attacks of rapid pulse, pain in the stomach and intestines, indigestion, dyspepsia, and urinary frequency. They complain of disorders of sex functions, of weakness and exhaustion. They are restless, excitable, and unbalanced. They are the product of the disappointments, fears, and hates of modern urban life. Their misfortunes are exacting a terrific toll in dollars of public money for their care and a more alarming toll in loss of mentally healthy citizens.

Hand in hand with frustration comes true insanity. Almost any investigation of modern life leads to the conclusion that the white man, with his machines and his cities, is definitely on the road to insanity. While the Selective Service recognizes many reasons for rejection, reliable authorities now claim that the nearer one lives to the center of a metropolitan area, the more likely he is to become insane. A recent study made in Chicago deals with the place of residence of those who have become insane during a particular period. It reveals that the rate of insanity is twice as great for those persons living in the center of metropolitan Chicago as it is for those living in the suburban areas.

It has been shown that insanity is not directly related to the economic status of its victims. Proof lies in the fact that there was little change in insanity rates during the depression. However, there may be a correlation in some instances. For example, Selective Service figures do show that in the New England states there is a connection between the rate of insanity and the economic level. This fact may be explained, in my opinion, by the fact that the errors of the breaking mind are reasonably certain to be reflected by an inability to get along in the world.

Another striking fact brought out in recent studies is that alcoholism is directly related to the density of population and not to the actual type of housing in which alcoholic victims live. Apparently crowded living conditions tend to make the drinker indulge to excess.

It is interesting to note that not all poverty-stricken areas produce the same kind of insanity. Different parts of a city, indeed, produce varied forms. A Chicago study, to which reference has been made, revealed, for instance, that the blighted rooming-house areas produce one type of insanity—paranoid schizophrenia—which is characterized by almost total indifference to one's surroundings. In areas inhabited largely by foreign-born, the catatonic type of schizophrenia predominates. In the areas inhabited by Negroes, dementia paralytica is the most frequent type encountered.

Worse than the curse of frustration, which is visited upon so many of the city's youth, is constitutional inadequacy, bred in its vitiated atmosphere, nourished in its unhealthful environment, and brought to fruition by maternal overprotection. Waiting in the office of every physician in the big cities are dozens of nervous, immature, unstable children, now grown to adulthood. They are breaking under loads which are too great for them to bear. Just as with the frustrated, these who are inadequate for the realities of life are trying to explain their failures to themselves, their families, and their friends on the basis of a set of complex symptoms which lead to the waste of much medical time and laboratory expense.

We physicians, in our attempts to help these people, too often succeed only in raising greater fears. After a visit to us and talking to patients in our waiting rooms, they worry about new and heretofore unthought-of diseases, for they are very much open to suggestion. All of this anxiety—added to childhood fears which have never been overcome, devastating experience suffered during adolescence, resentment against nature for real or fancied limitations, or a misadventure in vocational effort—poses a problem that is not difficult to recognize but almost impossible to solve. Many of these difficulties have arisen, of course, from the hurry and bustle of city life.

The background of personality often defeats the physician when he undertakes to treat a patient with some chronic disease. Personalities of this sort are common among the disappointed and the inadequate among the inhabitants of our cities and are not common among those who have a plot of ground to till. Sooner or later the physician awakens to the fact that his patient does not wish to get well; that he is getting all the pleasure he has in this world from being sick. This is the death-wish drawn out into a constant pain sensuality.

It is almost impossible for a woman to be a good and wise mother in the city. Unfortunately, too, the traits of childhood persist into adult life. In the city there are so many dangers that a child needs relentless watching by day and by night. Thus, in his early years, the mother sensitizes the city child's mind through overprotection. Such overprotection is not so bad when the mother concerns herself with the physical factors such as cold air, drafts, and infections; but in her desire to protect her offspring from the cruel happenings of life, of which her urban existence has made her conscious, she proceeds to protect her child from having to make decisions. Protection from physical hazards may damn the child's soul through frustration and the consequent introspection. He must be allowed to do the things that other children do, play their games, and share their experiences.

It is regrettable that mother love in its highest ecstasy blights and too often crushes the soul of the child with a cruelty that the mother never for a moment intended or even realized. This reaction bears no relation to intelligence or education. The fact is that this mother has overreacted to her situation—her small family, her cramped living quarters, and the artificiality of the urban life which engulfs her.

Every child needs rivals of its own sex and near its own age. Children need time to play. But the city child's schedule is a heavy one. What with his school work, his homework, and his music and dancing lessons, the child of the more fortunate couple has little time and often no place to play.

When summer comes, there is no relief. Like any other young animal, he should be turned out-of-doors so that his energies can be replenished for the coming winter. Summer camps are often too well run for him. Their regimentation of time and play serves only to draw him into himself instead of bringing him out of himself to face the facts of life. A better way to plan the summer for him would be to arrange for him to spend his vacation in a rural village or, still better, on a farm where he can learn to play with other children his own age—children whose point of view is not so sophisticated as that of his family and friends at home. Of course, the right thing to do would be to recognize all this as a compromise and to rear children away from the city in a rural environment, for the city is not a fit place for any young animal, least of all a child.

Frustration comes at every turn and with it, as a natural accompaniment, a sense of insecurity. Frustration is not peculiar to our social order, but it is common to the city life of every social order.

Disappointment with his job, his fellows, his family, and himself meets a person at every turn in the city. The larger the city, the more and the greater are the frustrations, for in a large city there are fewer opportunities for the individual to regain his composure, consolidate his position, and make proper social adjustments. With every disappointment he develops a clearer realization of his own insecurity. It breeds even in the most realistic heart a longing for security or, in more modern terms, for freedom from want, from fear, from sickness and old age. Complete security will never be possible to attain.

The factor that contributes most to the fears of the city

worker is the state of his finances. He has to take money from his pay for every single thing that he and his dependents get —every item of food and merchandise and service. This makes it well-nigh impossible for him to save towards a rainy day. His plight is like that of the one-crop farmer. He has no security. He almost goes out of his mind whenever he thinks of the hazards of life—sickness, death, old age, and, above all, unemployment. He may help start a revolution, but he can never improve his lot until he gets on a small tract of land big enough to support himself and his family. If he does not avoid the fear of insecurity, he must become a prey to the neuroses all too common in our time, but the opinion may be ventured, not without foundation in fact, that the security obtainable in small communities and rural areas is greater than in metropolitan centers. The social and economic proofs may be left to the authors of other chapters in this volume. From a medical point of view, however, that way of life which reduces the fear of insecurity to a minimum, without jeopardizing freedom, is highly to be desired.

Most illnesses can be traced to a poor state of nutrition. In recent years, some people have been greatly concerned with what they call adequate medical care for the underprivileged, but however much pills and potions relieve suffering and at times conquer infection, in many cases proper nutrition—the consumption of foods rich in nutrients, properly balanced—would have prevented the sickness.

As an illustration, witness the studies which have been made of Negroes in the slum areas of Cincinnati, where food supply, food preservation, and food habits are all bad. These Negroes die like flies from rheumatic fever, tuberculosis, and the diseases commonly brought on by overcrowding, bad nutrition, and the consequent lowered resistance. In a Negro suburban village near by, under average suburban standards of living, the relatives of the same Negroes die at the same rate and with the same diseases as do the white persons of the neighboring towns.

Very few persons are particularly interested in the health problems arising from poor food, poor ventilation, and crowding in the case of an individual colored citizen in these slums. However, once malnutrition has lowered the Negro's resistance and he has contracted tuberculosis, socially-minded citizens are greatly concerned about the adequacy of his medical care. Too often it is too late. Less crowded housing and the farm crops that were not raised and the milk that was not produced, if he had received them, would have prevented his having had tuberculosis in the first place.

Or again, during the late depression, a state dental society offered to fill, free of charge, the cavities in the teeth of all the children whose parents were on relief. This, too, was a noble gesture on the part of these dentists, but the cavities continued faster than the dentists could fill them. The reason is obvious. The children were not getting foods rich in calcium and vitamin D. It is, therefore, to be hoped that our attention will not be fixed on some scheme for getting sick people to physicians so much as on realizing that the people will be sick if they are not given proper nourishment. Prevention is the solution of the problem.

The history of Columbus, Ohio, offers an excellent example of the value of prevention. Some forty years ago in Columbus hundreds of persons were stricken with typhoid fever every summer. The physicians of the city literally lived off their income from typhoid fever. It was their "cash crop." The citizens determined to do something about the situation, and, after much discussion and many mass meetings, obtained a pure water supply for the city. Typhoid fever presently disappeared.

It is to be hoped that America is at the threshold of an epoch when it becomes the duty of society as a matter of public health and welfare to see to it that all of its members are nourished according to the scientific standards set for an "optimum diet." This does not mean doling out vitamin pills. It means seeing to it that American foods are made rich in the essential elements of nutrition through proper cultivation. It means, too, seeing that all people can get these foods. Finally, there must be a change in the sense of values which many of our people have. This means that many people must choose between some of our so-called modern luxuries and good food.

In the face of previous experience, it seems that the best way for the industrial worker to get the proper food is for him to go to a small five-or-ten-acre tract where he can raise almost all of the food for his family. Of course, along with this way of life must come education in proper food values. Food-farming must become a way of life for many of our people instead of a way of making money for a few. It likewise must become the answer to the individual worker's problem of what to do with his leisure time. With a backlog of suburban farming, he can afford to cut his work week short. In addition to spreading work sufficiently to allow full employment at nearly all times, this plan would give the worker and his family independence from unemployment, robust health, happiness, reasonable freedom from disease, and a sense of ownership in our country which would produce in each individual the highest type of national pride and patriotism.

I do not intend to extol the diet of those who live in the country above that of those who live in the city. Rural dietaries have been shown by many surveys in recent years to be inadequate. But they are inadequate either because the farmer tries to ape the bad food habits of his city cousin or because he needs education in the value of proper diet. The remedy is always at hand.

Nutrition is not a difficult subject to understand. Good nutrition depends upon good food, and good food depends upon the quality of the soil in which it is grown. The man who lives close to the soil, therefore, can with proper instruction get an adequate diet. As yet this is not possible for all the people in the city. The congestion in urban centers has produced problems in production and distribution which have not been solved. As a result, the greatest deficiency in urban nutrition is the lack of minerals and vitamins.

It should be understood that vitamins are not foods. They do not supply us with fuel, nor do they enter into the structure of our bodies. They are substances which aid us to burn up the foods we eat without overheating our bodies. Nature has put just the right amount of them in each mouthful of natural foodstuffs. This is the reason that it is harmful to depend upon white flour, refined sugars, and syrups for our food supply. It is unnatural and inefficient to take our vitamins in artificial and concentrated forms. Yet our city drug stores are selling these vitamins to two out of every five persons who enter them. These capsules are not meant for well people; they are only necessary for the physician to use with those who are sick. Persons with certain diseases which prevent them from absorbing and using their foods, those who must be on restricted diets, and those who have practiced bad eating habits so long that they have fallen ill of deficiency diseases need these supplementary vitamins. Well people should always get their vitamins in their natural state in foodstuffs from the grocery or, even better, from their own gardens.

Evidence is abundant that much damage has been done to the social, mental, and physical health of our people through the decline in the quality of our nutrition. Overrefinement of foods in the attempt to preserve them for city consumption has brought on many bad food habits. This has been the curse of our city food for a long time—processing and preserving for convenient packaging and a minimum of loss in distribution and storage.

In this way the city dweller gets what he wants, but he pays the most for the least. It is convenience and not nutrition that he is buying. Grains are ground, polished, bolted, bleached, and precooked, so that they may be dispensed without fear of spoiling—they are too devitalized to spoil. Wheat flour has been refined into a pure starch. As a war measure, the American people use so-called enriched white flour. This is not a natural food, but the same old starch to which a few minerals and vitamins have been added. Thus a precedent is established which may do much harm. Putting in a few synthetic substitutes for the many natural substances which were removed in processing is by no means true enrichment.

Vegetables are bleached; and this action destroys their chlorophyl and often impairs their mineral and vitamin content. Fruits and berries are cooked, canned, dried, treated with sulphur, or combined as a flavoring with large quantities of sugar to make jams, jellies, and preserves. While canning is necessary for winter use, the city dweller too often "lives out of cans." Citrus fruits are dyed or subjected to gas treatment to make the city resident in the north think they are tree ripened. Cane and beets are overrefined by taking out all of the vitamins and minerals to make a "pure" granulated sugar. Meats are pickled, soaked in artificial smoke solutions, needled, and injected with brine, gelatin, and liquid smoke. Now the city palate is tempted with tenderized meat obtained by allowing it to rot for a while and then stopping the putrefactive process. Eggs produced for the city trade must not be fertile. They do keep better, but they lack an essential hormone.

Even pasteurization, desirable as that may be when milk is purchased from many sources, destroys some of the vital substances in the milk. Thus the man who can drink milk from his own clean, well-nourished, healthy cow gets a good deal for himself and his children that the city family never has.

In the past, medical science has conferred upon those persons who have availed themselves of its knowledge the prevention of infections through vaccination and immunization, seemingly miraculous cures through new and powerful chemicals, and additional years through the better feeding and care of babies. In the future medical science promises to those individuals who will take advantage of the newer knowledge of nutrition and physical fitness greater stature, better physique, increased resistance to infection, comparative freedom from the so-called degenerative diseases, the insertion of at least ten vigorous years into the midspan of their lives, happier dispositions, and a higher level of cultural attainment.

These blessings, however, cannot come in any full measure to him who lives in the city. In any attempt to develop sound minds in sound bodies, the modern city stands indicted as an unfit place to live and rear a family. Our cities must be decentralized so that our people may be redistributed where they can be healthy and happy.

The task of getting our citizenry out of the restricting influence of city life, with its undue numbers of mental breakdowns and cases of insanity, its increased mortality, and its diminishing birth rate, demands immediate attention. The future of an increasingly large number of our people lies in a combination of urban and rural life. Our people in the cities are not able to face and solve their problems. The solution of most of our pressing health problems, as well as many of our economic problems, can be found in a new independence for industrial workers, when they have their wages as a "cash crop" while they raise a living for themselves and their dependents on a tract of land situated on a good highway reasonably close to their work.

This manner of living will bring genuine independence.

It will provide fresh air, light, sunshine, fresh nutritious food, real security against unemployment and old age, a greatly elevated standard of living, freedom from disease, dynamic health, and a busy, happy life for every member of the family. When this takes place, the city will no longer present any formidable biological or health problems.

6. *An Architect Protests*

HENRY L. KAMPHOEFNER

BY now it should be clear to the reader that this book is not concerned with the building of more cities, nor with the salvaging of those that have grown without plan or with plans that came too late. It is concerned, rather, with the best possible utilization of the broad acres of our land. In the vast intervals between the great cities of America, what can be done? And, more importantly from the architect's point of view, how can it be done to better the social and cultural pattern of the nation?

The Industrial Revolution, an expanding commercial economy, and two world wars have changed the fundamental structure of our civilization within the space of a century. Sound concepts of architecture and community planning have not been absent during the period, but they have often been frustrated, not only by the momentum of community growth, but also by the resistance and impatience of a people eager for "settlement" and profits. But it would be folly to permit some of the trends that have brought us to our present position to continue. It is not merely the unaesthetic that gives offense, but the social condition. Thomas Wolfe has something to say of it in his *Of Time and the River:*

"It was fabulous and incredible, but there it was. He saw again the million faces, the faces, dark, dingy, driven, harried and corrupt, the faces stamped with all the familiar markings of suspicion and mistrust, cunning, contriving, and hard and stupid cynicism. There were the faces, thin and febrile, of the taxi drivers, the faces cunning, sly, and toxic with unnatural fires. . . . The brutal, heavy figures of the Irish cops, and their beefy faces, filled with the stupid, swift, and choleric menaces of privilege and power, shining forth terribly with an almost perverse and sanguinary vitality and strength among the swarming tides of grey-faced people. They were all there as he remembered them—a race, mongrel, dark, and feverish, swarming along forever on pavements, moving in tune to that vast central energy, filled with the city's life, as with a general and dynamic fluid."[1]

This may be the observation of a rebellious and oversensitive spirit, but many readers insist that Wolfe does not exaggerate but merely portrays a spiritual blight to which the average city dweller has, unfortunately, become hardened.

Winston Churchill commented in British Parliament not long ago that, "We shape our buildings and afterward our buildings shape us." It would be easy to place the blame for the chaotic congestion of the city upon the skyscraper. However, it is not the skyscraper itself that is bad, but the relation of that great building to all other factors that surround it. Life in the city is characterized by dirt, noise, and confusion. The skyscraper represents an attempt to create more space in the midst of these conditions, but it adds darkness and congestion to the dirt, noise, and hurry of the streets that surround it. It aggravates the main evil of the city to the extent that it brings too many people into too small an area, thus withdrawing further from the ease, comfort, and quiet which modern man needs.

An architecture which fulfills our hitherto vague aspirations for space, repose, fresh air, and growing things in a community unspoiled by the helter-skelter packing of unrelated dwellings, business houses, boulevards, and traffic arteries must be based on the principles of city and regional planning as stated by such writers and critics as Mumford, Wright, Saarinen, and Hilberseimer. Proper regional and even city planning cannot progress, and no satisfactory solu-

[1] Thomas Wolfe, *Of Time and the River* (New York, Charles Scribner's Sons, 1939), page 416.

tion to overcrowding can be developed, unless our people will concern themselves not only with their own needs but with the needs of their neighbors, and, as the organic pattern develops, with the needs of their neighbors' neighbors.

Hence the desirability of some wider appreciation by Americans of the real function of architecture and of community planning. The architect's job is to conceive his building, if he thinks in broad terms, as the combination of all of the requirements of his time. He will apply the basic principles of art as an organic whole, and he will give to his structure a harmony in all of its parts, so that it will meet not merely the requirements of art but of life. As he follows these principles, he attempts nothing less than to integrate the expression of his own day with the continuum of all times.

Basic to the thinking of the architect as well as to the community planner must be the realization that, with an expanding means of transportation, man can now move rapidly, inexpensively, and efficiently. No longer is it necessary to conserve space—to move vertically as a device against time-consuming hauls horizontally. Decentralization, which was handicapped by transportation in the earlier phases of the Industrial Revolution, today has the means for accomplishing its objectives. Yet if decentralization is to correct the evils of the past, it must be organic. The mere multiplication of subdivisions, conceived with an eye to profit and dressed up with the slogans of repose, will not do. Community and regional planning must be carefully integrated over broad areas to avoid waste, duplication, and the back-and-forth haul. Transportation is a means to an end, not an end in itself. Ideally, its purpose is not to foster more and longer-distance commuting, but to eliminate overcrowding.

Social trends usually begin with the upper classes, and it is a commonplace that the more affluent members of society have been moving away from the city to the rural areas where they may have privacy, space, and peace of mind in natural and restful surroundings. The great city mansions of the late nineteenth century are being closed and abandoned. The country home has become the ideal for the city worker who can afford it. This movement towards decentralization of living and the return to the rural life is by no means a retreat into the isolation and the hearty and arduous deprivation that we have ordinarily associated with nineteenth-century farm life. The increased facilities for transportation and communication (such as radio and telephone) and the increase in labor-saving devices in the home have brought the advantages of cosmopolitan culture back to the country.

If we may assume that the *mores* and ideas of the wealthier classes are taken over by the middle and lower classes as fast as the latter can understand and afford them, it is an obvious corollary that the ideal of living just described will shortly be shared, at least as an aspiration, by the masses of our own society. Now, as it becomes apparent that the advantages of the simple life may be enjoyed along with the attractions of urban living, this rather vague ideal becomes translated into a tangible goal. An architecture based upon the principles already described in this chapter will provide the proper environment for such a life.

The idea of decentralization is to so order our society that these advantages will be possible for all economic and social levels of society. This idea is to create cities that will not be cities as we now know them, but to provide a new adjustment of man's social and economic activities to the land. The idea, obviously, is to eliminate the cities as centers and bring rural living and industrial activity into a more intimate juxtaposition.

Several methods for creating this new environment have been suggested by city planning authorities. At least two schools of thought are represented, both of which presuppose a fundamental replanning of the city. The first adopts the centric system and is now represented best in modern writing by the plans for satellite towns or communities. Frank Lloyd Wright's famous "Broadacre City" presents the nucleus for the second type of replanning, which has been called the ribbon-type city.

The satellite towns as a solution to problems of the city were first introduced by Raymond Unwin, the British city planner. His studies grew out of ideas developed by Ebenezer Howard as shown in his "Garden Cities."

Le Corbusier's *Une Ville Contemporaine* is based on the centric system of planning and is one of the best solutions for the larger communities. Le Corbusier's ideas are to concentrate the congestion of the residential areas into great skyscrapers. The concentration of the family units into great apartments then would provide space around each building for light, air, open spaces, and parks. The center of the city would contain the pivot points of the traffic system at the central transportation center and would be comprised of other great skyscraper buildings containing from twenty thousand to fifty thousand persons each, or a total of about half a million persons in the business area. The residential area would surround the business section and would be in units as described above. At a distance from the residential area would be the industrial area, separated from it and permanently zoned as a park or green area. In the surrounding country would be the suburbs arranged as garden cities in the satellite system, with permanently restricted green areas for separation spacing. Le Corbusier shows the practical application of this system of planning in his study for the replanning of Paris. As the town or city folk move out in the direction of greater space, subdivisions or satellite communities should be planned to take an optimum of not over one thousand families. In order that selfish real-estate interests might not mutilate the green belts so needed around the periphery of such areas, the belts would be permanently zoned and city owned for no other purpose than for park and play areas, always green and inviting. The city dump and junk yards should be placed in their most advantageous position for the good of all and for the protection of health standards.

Frank Lloyd Wright has corrected many of the weaknesses of the centric plan of the satellite town in his studies for the ribbon-type Broadacre City. The ribbon-type city is also masterfully presented by L. Hilberseimer in his recent book, *The New City*, a scholarly synthesis of the best thought on the subject. Professor Hilberseimer's preoccupation with the smoke problem, even for a Chicago dweller, at times seems all-consuming. That problem might better be solved by the Broadacre City principles, through which all coal is burned at the mines and transmitted as net power, thus eliminating transportation problems and contributory factors to the ugliness and unhealthful conditions in communities where coal is burned.

Hilberseimer's ribbon-type city has been developed by the designer as a planning system based on the independent settlement unit which is of a limited size and contains within itself all the necessary elements of a city segregated according to function. The entire town in this city type is built around the main arterial highway or traffic artery. On the side of the highway in the direction of the prevailing wind are all elements of the city that must be free from the nuisances of noise and smoke. Against the highway and between it and the residential area would be the commercial and administrative buildings, set within a green belt to give proper light and air. Beyond the commercial section toward the breeze would be the residential areas with their accompanying schools, recreation areas, and community buildings. An agricultural area would adjoin the park. On the opposite side of the arterial highway, where the prevailing breezes carry the noise and smoke nuisance away from the residential section, would be the industrial areas of the city. Smokeless and quiet industry could be developed along a line between the highway and the heavy industry, and would actually form a cushion of space between the living areas and heavy industry. The residential areas would be within walking distance from the commercial areas and also from the industrial or general working areas. The residential groups would be related to and connected with each other by a simplified traffic system. Because of the flexibility of this plan along the traffic artery, the city in which it is used might be large or small, it might increase or decrease in size, but it would always remain a working entity. Each group within the city could be extended or reduced in size without disturbing the life of the city as a whole. Open spaces between the groups could be used for gardening and farming as the city would be reduced in density as it stretched toward the open country. Garden and farming areas so disposed between the residential, commercial, and industrial areas would achieve an organic relation between industry, commerce, and agriculture. Rail, bus, and automobile traffic would be simplified and made more efficient in such a rationally planned community. The accessibility to means of transportation could be practically the same for all persons living in the community.

It can be seen that this city type can be extended and reduced along its main traffic core, also that new cities can be developed in the great regional plan in relation to the dispersion and the disintegration of their particular smoke nuisance. Therefore new cities of the ribbon type will run approximately parallel to the first laid out cities of the type, with variations in planned location dictated by topography and natural resources. Planning such a city means the restudy and re-evaluation of all parts of the present community, and not just the cutting away of an isolated diseased part.

A better choice would probably conform precisely to neither of the two general plans, but would draw upon the best features of both. Such a community would number from five thousand to twenty thousand inhabitants—places big enough to contain the necessary economic drives and opportunities, yet small enough to permit a healthy relationship between industry and commerce, on the one hand, and living, on the other. Such communities could fit into the vast intervals between our great centers of population. Within these areas, often close to the raw materials and transportation, are small communities which today are the embryos of future great agglomerations of population. If the trends of the past continue and are abetted by boosterism, new cities, even vaster perhaps than the old, may arise. But such a development does not serve the ultimate purpose of society; the community where life finds its best and most natural expression will be neither vast nor towering.

In the smaller community, there would be a sound balance between production from the land and production from factories. Those who dwelled there would indeed have one foot on the land: space enough so that all who wanted it might work from one to five or ten acres. Certainly all could avoid the jam-packing that is now the order of the day. By careful planning, even factories and shops could be placed in near-rural surroundings. Shopping areas there would be, but not ganged to one long Main Street. Rather they would be dispersed throughout the residential areas according to need. Such a decentralization would provide convenience for the community and greatly reduce the traffic problems normally inherent in community life. If transportation is used merely to add to the hurly-burly of existence, it has not accomplished the objective which is really being sought, namely, to assure quiet and restful conditions.

The organization of a community requires a functional approach. If there is very little of such thinking now, we may charge it up to the force of traditionalism. For example, in the small communities that have been expanding perceptibly within the past decade, precisely the same trends are observable that we have witnessed in large cities. Instead of bringing organic principles to bear, residents allow crowding to take place at the center, permit serious traffic problems to grow, and then create for themselves residential areas on the periphery of the community. It is a vicious circle.

Organic planning calls for the convenient, harmonious location of each element that goes into the building of a community. Not a mile-long Main Street with shops, but decentralized shopping areas; not a factory district with noise, confusion, and dirt, but a factory on broad acres with attractive homes, again on broad acres, within convenient distance; not a super traffic artery as a palliative for bad planning, but many quiet streets.

Some of these desirable features may be seen in Greenbelt, Maryland, which in a period of twelve years has managed to adjust factory production, shopping areas, quiet residential areas, and space to the requirements of sound community development.

From a more strictly architectural point of view, what course should a decentralized community take? In a sense, the job begins with one resident building his house in relation to and in consideration of not only his own needs but the best needs of his neighbors. Such a community interest can be extended to the business, industrial, and shopping areas, so that designs may be co-ordinated into harmonious groups of buildings rather than the present hodge-podge to be seen in any community, large or small.

At the outset, a basic consideration—orientation—would be possible, whereas it cannot be in the chaotic congestion of the metropolis. It is a pathetic tribute to the architecture of America that so few houses can be lived in through such summers as we have had in recent years. Surely it is not a good house if the occupants must spend their evenings in the public gaze in the front yard and their nights on a mattress in the back yard. The nature of the climate, that is, the direction and path of the sun through the four seasons of the year, the prevailing breezes, and precipitation averages, should influence the character and spirit of the house. Astronomy plotted the rhythmical rotation of the sun centuries ago. Its exact position in the sky and its location on any future day can be forecast with infinite accuracy. Modern meteorology has carefully recorded wind direction and velocities, and a *wind rose* is available for all localities in the country. Such information will show, for instance, that in central Oklahoma some 80 per cent of the breezes blow in the southeast-southwest quadrant through the six hot summer months. Why is it, then, that architects, having this information available, have not used it in working out the controlling systems of their designs. With more than optimum concentration in almost all of our large cities and in many of our larger towns, orientation, which is an integral part of intelligent site planning, is almost impossible. With the decentralization of living now made possible, adequate space and carefully studied orientation should be more than just dreams.

As long ago as the first century Vitruvius, the great writer on architecture, stated that each room should be planned for its specific use, with exactly the correct orientation to gain sun and warmth, or coolness, as the case might be, to give the proper views, and to have the simplest connections to other rooms with which it is related. Such a statement made by that influential writer nearly two thousand years ago makes it appear that orientation and proper consideration for climatic conditions is hardly a radical idea. Vitruvius did exert tremendous influence on architecture, at least up to and through the Renaissance. However, his ideas on orientation could not be followed in the present day because of the congestion of cities. If intelligent regional planning gives us either the satellite or the ribbon-type community, minimum congestion for the dwelling can be a basic consideration. The decentralized community can be conditioned to the realities of nature.

In the great architectures of the past, most notably in classical and medieval times, the expression of the nature and the purpose of materials in building gave one of the greatest single qualities to the architecture of the periods. That phrase, "the nature and the purpose of materials," should give the contemporary architect the key and bring to modern man one of the most vital and important considerations in the development of an organic architecture.

It is a significant fact that many of the finest creative and expressive uses of materials are reflected in the rural areas of America, for example, in the Pennsylvania barns and farmhouses or in the great grain elevators and silos of the West and Middle West. There materials express their true integrity. While in the cities, where the hypocritical life of ostentation flourishes, we find soft woods used to imitate harder woods, metals to imitate something else, plastics to imitate metals and even cloth. Imitation has become the standard expression of modern city life. We should take great care that such an attitude does not spread more than it has to the rural areas of our country.

The modern farmhouse and its accompanying barns and outbuildings, if planned properly, can be simple and direct in their relation to their natural settings and to the farm families and the animals for which they provide shelter. Modern building materials and techniques combined with contemporary principles of organic planning can now be used in the development of the farm group to increase the standard of living of the farm to a level much higher than is now found in the urban community. The modern farm group will be a planned and unified organization, conditioned to its natural setting and the life of the living organism for which it provides.

A number of reasons have already been given for the failure of cities to contribute adequately to the cultural development of the nation. One of the greatest retarding factors has been the blind following of transplanted and false traditions or adherence to "style." Eliel Saarinen, the architect-director of the famous Cranbrook Academy of Art, discusses the effect of blind eclecticism on our architecture from an interesting and critical point of view in his book, *The City*,[2] from which the following passages are taken:

"Fully to grasp the momentousness of the step toward imitation we may suggest historical parallels:

"Supposing—to begin with a rather absurd thought—that the Greeks had adopted the already ripe Egyptian form. In such a case the Greek cultural form would not have come into existence, but the Greeks would have been compelled to dwell midst an artificially superimposed style form. Supposing furthermore—to continue with another absurd thought—that the Medieval time had not had the ambition to create its own form, but had imitated the Greek one, or any other previous style form: there would not have been born Romanesque or Gothic art, but only a dead Medieval decoration borrowed from somewhere else and, therefore, lacking cultural significance. Surely, in both of these instances, the Greek and the Medieval, such an imitative action would have been much the same as cultural suicide. This statement is plain and clear to everybody, and does not call for further comments. And even if some Greek architect had built only a single Egyptian portico midst the Greek form-world, or some Gothic builder had erected only a single Greek temple in the Medieval town, even those incidents no doubt would be regarded, today, by everyone, as stupid doings.

"Well, on the other hand, look at the towns and cities from yesterday. There one can find more than plenty of worthless imitation of Greek, Roman, Byzantine, Romanesque, and Gothic style. There one can find all the Renaissances, general and provincial. One can find all kinds of

[2] Eliel Saarinen, *The City*, (New York, Reinhold Publishing Company, 1943), pages 76–77.

ramifications and bastards of almost any style that has been created by previous eras under this sun. Moreover, one can find that these styles have been mingled together without any logical reason whatsoever, and one can soon discover that the architects, speaking generally, have selected their styles just as easily and capriciously as one selects the pattern and color of one's shirt and tie."

In the course of time it has been forgotten that architecture—just as any other art form—is fundamentally creative and that style evolves gradually from the progress of creating architecture, and not architecture from style. As these new communities of which we have been speaking begin to grow organically in the rural areas of America, a great style will emerge from them. But as long as we continue to base our art forms on eclecticism and false tradition, our cities will continue in their present chaos.

If architecture is to keep pace with a changing society it must be approached dynamically. The great main façade of the cathedral at Chartres illustrates a dynamic approach to architecture. When the time came to add the second great tower, architectural expression had changed since the days of the original building. Rather than consider the creative expression of architecture a static thing, the designers built the second tower with a contemporary expression, using similar materials in a harmonious blending, but with a modern character and expression. Architectural expression then kept pace with the development of its art form.

By and large, twentieth century architecture has been static. It has clung to false and little understood traditions and expressions of the past while trying to solve contemporary problems. The architect who yields to the pressures of a client who seeks solutions based on ignorance and misunderstanding, rather than to use the influence of his superior experience and training to persuade the client to do what is architecturally sound, prostitutes his profession with a kind of mongering. There is no evidence that an eclectic or ready-made architecture has obtained any great past results or that it has anything to do with our life as we live it today. If our architecture is to become great, it must proceed from the ground, it must be conditioned to the terrain, it must recognize the native industrial and social conditions and provide for them, and it must acknowledge the nature and the purpose of the materials with which it works. These factors must inevitably determine the form of the buildings we build.

If modern architecture in our new cities has a dynamic future as compared to a revival or even a passive form, it must become an integral part of our social thinking. It must be organic. It must discard the blatant ostentations and superficialities of the present city. It must spread out and demand space for the contemplative development of man. Human nature must be one of the materials served by the building and human nature must serve it. As has already been said, such an architecture is, and as a matter of course must be, an actual interpretation of social man. It must co-operate with nature rather than work in opposition to it.

Victor Hugo, who was regarded as a great modernist in his day, stated prophetically in a chapter in *Notre Dame:* "Architecture, already some five hundred years moribund, will in the latter end of the nineteenth century or the beginning of the twentieth, come alive again." When the architects and their clients recognize the requirements of nature and the needs of human beings in creating forms for their buildings, that prophecy will come true. When the first imperfect Broadacre City or ribbon-type city has become a physical reality, it is reasonable to suppose that the new way of life which it will support will exercise a profound influence on the attitudes of its inhabitants. Space, integration of building and terrain, and the charm of rural living will become as compelling ideals and as engrossing objects of civic pride as the skyscraper used to be. When this happens, there will be a renaissance of general creative interest in architecture. People will no longer think in terms of single monumental buildings, but of buildings as parts of a larger, well-integrated community, and they will take as much pride in arrangement, relationship, and disposition of buildings as they used to take in the ostentation of a single edifice towering over acres of slums. Thus the ideal of order will replace the ideal of ostentation, and this new direction of men's thinking will lead to the creation of a genuine American architecture. Let us emphasize that a style cannot be invented or copied, it must grow out of a way of life. When the ideal of order is within men's grasp, architecture will flourish in America as it never has before. Thus a better life and a better architecture will promote each other step by step.

7. *Social Man and His Community*

J. J. RHYNE

IN any consideration of the social aspects of modern life, the complex culture pattern of the metropolitan community is obviously a phenomenon and a problem. The highly artificial life of the great city, the stimulus which comes from close association of vast numbers, the excitement and strain under which many in the large cities live, and the highly developed competitive struggle to maintain status have combined to give the great city a disproportionate share of the major ills of human society. The fact that the metropolis furnishes a great attraction to the average individual in general and to those tending toward the abnormal in particular is perhaps a sufficient reason for the undue concentration of social problems in the larger places.

For those who are average and normal, the hazards of metropolitan life are not escapable. Inevitably many such individuals, thrown into the keen economic and social competition of the great city, find it impossible to make the necessary social or economic adjustments. Despite the half-felt existence of these hazards, the healthy are attracted to city life because of its excitement and other inducements which are believed to be absent in small communities and rural areas. When adjustment fails, the resulting social problems are hardly less difficult of solution than those which develop for both native residents and unadjusted persons drawn from other areas.

The chances are that, along with the perfectly normal rural and small-town residents who have been flocking to the great cities for the past several years, an undue proportion of the less well adjusted have also found the attractions of city life irresistible. These persons, added to the native unadjusted products of metropolitan centers, have presented the social thinker with problems of the first order.

However, it should not be assumed that all of the problems of city life are induced by the unadjusted elements. The complex structure of the large city is in itself a very important factor, for this very complexity has made the development of social and economic controls difficult if not impossible. The progress of industry, the fluctuations in business activity, the technological changes caused by new inventions, and the economic and social dislocations resulting from these changes have thus far operated as "free forces," largely beyond the control of man. It hardly helps the situation to say that those who are caught by these forces are innocent victims. Rather it is to be hoped that means may yet be found in a free economy, operating in turn under a free society, to better the lot of the individual.

The social problems of the metropolitan community appear to fall into two general groups. One consists of those which are generated primarily by the operation of an industrial society; the second, of those that are more directly concerned with human beings and their behavior. By comparing the social problems of the great metropolitan centers with those prevailing in the less complex culture patterns of the smaller cities, towns, and the open country, the prospect of the individual's good may be made more real.

Man's three basic needs are food, shelter, and clothing. Of these, food is, of course, the primary essential, for without it neither man nor any other form of life can survive. Obviously, those who gain a livelihood by tilling the soil are least removed from the natural means to that end. Yet, as man has developed his culture, the ways of getting a living have grown increasingly more diversified and interdependent. Most of us today get no part of the food we eat through tilling the soil—the physical structure of our urban areas precludes any such possibility. On the contrary, we sell goods or services, and with the proceeds we purchase the things we need. The implications of this state of affairs become clear, apparently, only in periods of economic crisis.

Simply stated, the bulk of the population of our cities and towns is wholly dependent upon more or less regular jobs for the essentials and comforts of life. When the income from a job stops or is seriously reduced, even for a short period of time, the living habits of the urban dweller receive a serious shock. Adjustments must be made; the food budget must be reduced, cheaper rent must be sought, the outlay for new clothes may cease entirely. If the period of unemployment is sufficiently prolonged, resort must finally be had to a relief agency.

These contingencies are well understood by urban dwellers, most of whom assess the advantages of urban life in terms of comforts that the farm family is unable to afford. Farm life is characterized by long hours and heavy work, ordinarily for relatively small cash returns. The shorter hours of urban work, at what appear to be unusually high wages, have offered an irresistible appeal to many rural folk. During the past half a century, millions of people have come to the city because of this single advantage. In so doing, they have given up what is perhaps a simpler, though more dependable, means of livelihood in exchange for a more remunerative and a less certain source of income. Some have profited from the exchange of country for city abode; many thousands of others have been less fortunate.

Today, not one in five of us gets his living directly from tilling the soil. About one-fifth of us who do not get our living from the soil, however, are still resident in places of less than 2,500 population. In other words, nearly two of us in every five live either directly from the land or in such close proximity to the source of food supply that we would be able to make some adjustment should we lose our jobs or should our incomes be materially reduced. The flow back to the farm with the onset of every depression is proof of the fact that many of us return to the land when the job in town ceases.

The large part of that flow back to the farm comes from those who either are resident in the smaller places or have moved to larger towns but have relatives or other connections on the farm. In either instance, the shock of depression and the resulting loss of the job by those who have the opportunity to return to a farm is not so fraught with the elements of tragedy as in the case of the residents of our larger cities who are unable to make such adjustments. It is in connection with the unemployed in our large cities that we are brought face to face with the stark realization of the fact that our highly complex urban-industrial society has produced a relatively insecure source of livelihood for a large part of the population.

Because of the highly artificial life of the city, the cost of goods and services, and the complete dependence of the urban dweller upon a steady pay check, we may expect the shock of depression and the resulting unemployment to react more violently upon the population living in the large urban places. Because of the fact that the urban dweller, for example, is likely to be completely without the means of raising any portion of his own food, any loss of job or of income will have immediate and serious consequences for him and his family.

In times of depression it is obvious that the problem of unemployment first becomes acute in urban areas. In some cities the situation becomes serious before it does in other areas, depending in general on the nature of the industries in which the great majority of the people of a given city are employed. In general, it may be said that the residents of a one-industry metropolitan center are more subject to wholesale unemployment than the residents of a city with a wide variety of industries. Conversely, in prosperous times, the residents of a one-industry metropolis may be relatively better off than the residents of a multi-industry metropolis. Especially is this true when there is a heavy demand for the goods and services produced in the one-industry city.

Commercial cities, also, since they do not generally enjoy at any time the high level of prosperity sometimes enjoyed by the one-industry city, generally experience a rather gradual population increase. Conversely, the one-industry city is likely to grow by fits and starts. In one decade such a community may double its population, but in the next it may actually experience a decline.

The quality of artificiality in city life previously referred to is likely to develop in its most pronounced fashion in the one-industry metropolitan area that over a considerable period of time enjoys an unprecedented prosperity. It is when the inevitable period of readjustment comes to that type of community that the recuperative forces are least likely to be found effective. The whole community is likely to find itself in a state of economic collapse. It suffers much in the same manner as a plant which depends primarily on a single taproot for sustenance. When the taproot is severed, the whole plant wilts.

The city of multiple industries, or the commercial city, on the other hand, though perhaps never enjoying the highest peak of prosperity or experiencing the greatest rate of population growth, is likely to be able to withstand a greater degree of shock attendant upon economic dislocation than is the single-industry city. The city that grows more slowly has had a greater period of time to adjust itself to changing circumstances. It has had a greater period of time to develop naturally, as far as the forces of city life permit natural development. In times of depression it is seldom that all industries or businesses will be struck an equally hard blow. Since different levels of depression are generally experienced, some reach bottom ahead of others and begin to revive, to be followed later perhaps by others. All of the employment resources are almost never at rock bottom at the same time. That fact makes the recuperative powers of the many-industry and commercial city superior to the recuperative powers of the single-industry city.

In general and at best the large city is not able to sustain itself in times of severe economic stress as well as the smaller community or the open country. This fact remains whether the metropolis is fundamentally a single-industry, a multiple-industry, or a commercial community. Nevertheless, this differentiation on the basis of number of industries is significant in analyzing relief loads. In general the needs and the expenditures for relief in the city of many industries will be smaller per capita and in the sum total than the needs and the expenditures for relief in cities which must depend primarily upon the single industry. For the latter, when economic crisis occurs, the problem is all but insuperable; relief agencies are likely to find their resources overwhelmed from the sheer pressure of numbers needing care.

In a study of experienced workers on relief in March, 1935, it was found that "rural workers were underrepresented on relief registers, while urban workers were overrepresented."[1] Whereas urban workers in 1930 composed 60.9 per cent of all workers in the United States, urban workers on relief in March, 1935, composed 65.4 per cent of all workers on relief. Stated conversely, while rural workers in 1930 composed 39.1 per cent of all workers, rural workers on relief composed 34.6 per cent of all workers on relief. In other words, urban workers on relief were overrepresented by approximately 7.4 per cent in terms of all urban workers in the country in 1930. Although the overrepresentation of urban workers on relief was not, on the basis of this study,

[1] Division of Social Research, W.P.A., *Workers on Relief in the U. S.*, March, 1935, page 1.

particularly outstanding, nevertheless the degree of overrepresentation is worthy of further consideration.

In Table I is given the number of experienced workers on relief by states. It should be observed that New York, the most populous of the forty-eight states, had a smaller proportion of workers on relief than the number of workers in the state, thus indicating an underrepresentation. Pennsylvania, on the other hand, was overrepresented with workers on relief by approximately 28 per cent. Ohio also was overrepresented in the number of workers on relief by slightly over 18 per cent. The question may be asked, why were the states of Pennsylvania and Ohio overrepresented by workers on relief whereas New York was underrepresented? The answer may be found in the previous assumption regarding the sources of livelihood to be found in a given area. In states having relatively less diversification of business and industry, the chances are that in times of depression proportionately more of the workers will be forced on the relief rolls. This is a possible explanation of the overrepresentation of Pennsylvania and Ohio. In both states the iron and steel industries, along with the closely allied industry of coal mining, tend to dominate the picture. In New York state, on the other hand, is found perhaps the most extensive example in the country of diversification in business and industry, a condition that enables the population of that state, in one of the hardest years of a decade of depression, to be underrepresented on the relief rolls.

On the other hand, it would be erroneous to assume that in every case of a one-industry state there would be an overrepresentation of workers on relief. For example, in the case of Rhode Island, which is predominantly given over to the manufacturing of cotton textiles, there was a slight underrepresentation. The situation in the case of Rhode Island is even more significant when it is remembered that it is the most densely populated state in the United States.

By way of summary, the following comparatively highly industrialized states were underrepresented on the basis of the number of workers on relief: New York, Illinois, New Jersey, Michigan, Indiana, Wisconsin, Connecticut, Rhode Island, Maryland, and North Carolina. On the other hand, the following rather populous states were overrepresented on the basis of workers on relief: Pennsylvania, Ohio, Texas, California, Missouri, Minnesota, Georgia, and Kentucky. Massachusetts was neither underrepresented nor overrepresented on the basis of workers on relief. It will be observed that most of the highly industrial states of the Northeast had fewer workers on relief in proportion to all workers, whereas a number of the states, especially those centering around the Great Lakes, were overrepresented by workers on relief. A more thorough analysis of the types of industry in these two groups of states than time or space will permit here should reveal some very important facts and forces operating to provide a more secure economic base for the population of one state as compared with another.

An important case in point is the comparatively recently industrialized southern state of North Carolina. Whereas that state had 2.3 per cent of all the gainful workers in the country in 1930, only 1.8 per cent of all workers on relief were in that state. In other words, the rate of underrepresentation of workers on relief in that state was approximately 28 per cent. Cotton textiles, furniture, and cigarette manufacturing are the dominant industries in that state. North Carolina is one of the best examples that can be found of a state with a wide degree of decentralization of industry. In 1940 only one city in the state, Charlotte, had as many as 100,000 persons. By far the predominant part of the textile industry is located in towns under 10,000 population. Gastonia, the leading textile center of the state, had less than 30,000 persons in 1940. Many of the cotton mills are located

TABLE I[2]

Experienced Workers on Relief in March, 1935, and Gainful Workers in April, 1930, for the United States by States.

States	Number: Gainful Workers 1930	Number: Workers on Relief 1935	Per Cent: Gainful Workers 1930	Per Cent: Workers on Relief 1935
New York	5,294,305	581,136	11.5	11.3
Pennsylvania	3,538,203	507,710	7.7	9.8
Ohio	2,479,203	330,939	5.4	6.4
Illinois	3,045,790	328,866	6.6	6.3
Texas	2,067,123	248,375	4.5	4.8
California	2,381,312	240,118	5.2	4.6
Massachusetts	1,718,514	195,994	3.8	3.8
Missouri	1,355,792	185,740	3.0	3.6
New Jersey	1,640,599	169,861	3.6	3.3
Michigan	1,838,324	157,723	4.0	3.0
Oklahoma	781,846	152,490	1.7	2.9
Indiana	1,167,063	123,843	2.5	2.4
Minnesota	935,494	112,258	2.0	2.2
Wisconsin	1,064,866	105,367	2.3	2.0
Georgia	1,055,194	102,937	2.3	1.8
Kentucky	831,860	101,270	1.8	2.8
North Carolina	1,043,082	94,511	2.3	1.8
Kansas	649,664	90,714	1.4	1.7
Tennessee	883,644	88,504	1.9	1.7
West Virginia	538,927	75,090	1.2	1.4
South Dakota	233,437	72,542	0.5	1.4
Iowa	854,082	71,570	1.9	1.4
South Carolina	613,752	69,955	1.3	1.3
Louisiana	758,441	65,755	1.7	1.3
Virginia	818,818	65,678	1.8	1.3
Washington	622,387	64,439	1.8	1.3
Alabama	921,484	64,407	2.0	1.2
Colorado	375,930	64,060	0.8	1.2

[2] *Ibid.*, page 2.

States	Number: Gainful Workers 1930	Number: Workers on Relief 1935	Per Cent: Gainful Workers 1930	Per Cent: Workers on Relief 1935
Arkansas	605,129	62,091	1.3	1.2
Florida	562,667	59,360	1.2	1.1
Mississippi	740,540	57,486	1.6	1.1
North Dakota	227,116	54,638	0.5	1.1
Connecticut	641,976	51,902	1.4	1.0
Nebraska	479,510	50,612	1.1	1.0
Maryland	634,045	50,228	1.4	1.0
Oregon	391,580	38,875	0.8	0.7
Montana	203,140	30,014	0.4	0.6
Utah	160,767	28,436	0.4	0.5
New Mexico	133,016	28,186	0.3	0.5
Rhode Island	282,651	25,861	0.6	0.5
Idaho	151,357	23,201	0.3	0.4
Maine	282,855	22,091	0.6	0.4
Arizona	157,202	20,403	0.3	0.4
Dist. of Columbia	234,334	17,536	0.5	0.3
Wyoming	87,902	9,941	0.2	0.2
New Hampshire	177,872	9,397	0.4	0.2
Vermont	129,455	9,605	0.3	0.2
Delaware	91,745	5,900	0.2	0.1
Nevada	40,509	4,011	0.1	0.1

in strictly rural areas and answer best to the description of rural mill villages.

Because of an unfortunate combination of drought, dust storms, and depressed farm markets, some of the states with the greatest overrepresentation of workers on relief were located in the area more or less commonly referred to as the "bread basket" of the nation. Kansas, for example, had 1.4 per cent of all workers and 1.7 per cent of the workers on relief. North Dakota had 0.5 per cent of all workers, and 1.1 per cent of the workers on relief. Oklahoma had 1.7 per cent of the workers, and 2.9 per cent of the workers on relief. Montana had 0.4 per cent of the workers, and 0.6 per cent of the workers on relief. Other than the so-called "dust bowl" area, on the other hand, the predominantly rural states were generally underrepresented on the basis of workers on relief. Mississippi, which along with North Dakota is the most rural of the states, had 1.6 per cent of the workers and only 1.1 per cent of the workers on relief. Iowa, similarly, with 1.9 per cent of the workers, had only 1.4 per cent of the workers on relief.

The ratio of workers on relief to all gainful workers in twenty-five of the large cities of the country in general tends to bear out the thesis that in times of depression the city that is dominated by one or two basic industries generally has a more serious relief problem than the commercial city or the city with a wide variety of industries. Cities with an overrepresentation of workers on relief included Philadelphia, Cleveland, Pittsburgh, St. Louis, Boston, New Orleans, Buffalo, Cincinnati, Minneapolis, Toledo, Newark, Indianapolis, Columbus, Rochester, and Omaha. On the other hand, cities with an underrepresentation of workers on relief included New York, Chicago, Los Angeles, Baltimore, Detroit, San Francisco, and Washington. In general, those places which were underrepresented answer either the general description of a trade center or commercial city, or are characterized by a wide variety of industrial development.

A rather general substantiation of the validity of the figures taken from the preceding study is found in a comparative study of population and federal relief expenditures by states in 1933.[3] Of the more populous states, the following received a greater proportion of all relief funds than the population of the state warranted in terms of the total population of the United States: New York, Ohio, Illinois, Michigan, Wisconsin, Iowa. New York, with 10.3 per cent of the population, received 11.94 per cent of the relief funds. In the preceding study it will be remembered that New York was un-

[3] Odum, *Southern Regions*, page 454.

derrepresented on the basis of the number of workers on relief. Perhaps the most outstanding contrasts between the two studies are to be found in the cases of Pennsylvania and Michigan. In the study of population and relief funds, Pennsylvania, with 7.8 per cent of the population, received only 4.61 per cent of the relief funds. In the other study, Pennsylvania was rather seriously overrepresented in terms of the number of workers on relief. In the case of Michigan, a second instance of reversal of positions is found between the two studies. Whereas Michigan received more relief funds than the population warranted, 3.27 per cent of the population and 5.71 per cent of all relief funds, the state two years later was underrepresented. In the case of California, a reverse situation is also shown from the former study. In 1933 that state received 7.91 per cent of the relief funds while in 1930 it had only 4.6 per cent of the population of the country. Ohio, on the other hand, consistently showed a serious relief situation in both studies. In 1933, that state, with only 5.4 per cent of the population received 7.7 per cent of the relief funds.

Although the results of the two studies of the relief problem by states fail to harmonize in a number of instances, in general, it may be rather definitely concluded that, in times of economic stress, the population in the larger urban centers tends to become overrepresented on relief, whereas the population in the smaller places and in the open country is more likely to be able to meet the emergency without as widespread resort to the relief rolls. Since increasing industrialization is also accompanied by increasing hazards of economic insecurity, the logical conclusion would seem to point to the apparent fact that the greater the degree of concentration of the population in the large metropolitan centers the greater the degree of economic insecurity that concentration entails. The basic soundness of this conclusion is rather definitely indicated from the fact that within the same state, or geographic area, the proportion of the population that is being cared for by the social agencies during periods of depression in the larger cities is ordinarily much higher than the proportion of the total population similarly being cared for in the smaller cities and in the open country districts within the same area. This statement was true particularly during the last depression of the highly industrialized Middle Atlantic and New England States. In New York, Pennsylvania, Massachusetts, and New Jersey, for example, the expenditures during the depression in the 1930's for both direct relief and work relief were higher both totally and proportionally for the metropolitan areas of New York City, Buffalo, Philadelphia, Pittsburgh, and Boston than for the smaller places and the rural areas within the same states. During the period relief expenditures were at a minimum in the rural areas especially of central and western New York, central Pennsylvania, southern New Jersey, and western Massachusetts. Likewise, the relief needs of the relatively rural states of Maine, New Hampshire, and Vermont were proportionally smaller than for the large urban areas mentioned.

The only exceptions to the situation in which the larger places were overrepresented on relief when compared with the smaller places and with the open country areas during the depression decade of the 1930's were found in the sections of the country hardest hit by recurrent droughts and dust storms. In Oklahoma, for example, the relief expenditures for both Tulsa and Oklahoma City were relatively smaller in proportion to the population than were the relief expenditures in some of the strictly rural counties. The long-run effects of soil erosion and drought may be still further seen in the fact that the old age assistance rate continued abnormally high in a number of rural counties in eastern and southern Oklahoma not only during the depression period, but also during the abnormally prosperous period from 1941 to 1944.

Economic hazards, economic maladjustments—these are not merely ills in themselves: they produce social consequences which are nowhere more palpable than in the city.

It should be remembered that good citizens are made, not born; that criminals, like weeds, grow up without the benefits of a positive culture. Given a normal physique and mental capacity, accompanied by the usual natural urges and drives, the individual must have also the benefit of those positive forces which are necessary to the building of a law-abiding citizen. Otherwise we cannot expect him to develop into an adult with the proper respect for the rights of others. Social responsibility, unlike the measles, cannot be caught. It must be taught the individual and inculcated and integrated as part of his experiences over a long period of time.

None of us at birth, for that matter, regardless of the quality of the home life into which he may be born, is given an assurance that he will develop the degree of social responsibility required of a respected and self-respecting member of human society. But, given a good home environment, he is much more likely to acquire those traits and characteristics on which society has stamped its approval. When parents are themselves the products of cheap-rent districts, when they have not had the opportunity to develop normal human reactions to the social patterns and laws which society has laid down over a long period of time and as a result of a long train of human experiences, it is not surprising to find in the more densely populated cheap-rent districts a whole pattern of life that may be of a negative, if not of an actual antisocial nature.

When young people are born into that type of social world, when their social inheritance answers to the type of society just described, it is not surprising that the whole trend of a developing young life becomes warped and twisted emotionally and socially. Under such circumstances the developing individual is hardly given the opportunity to develop a normal, wholesome outlook on life. All of his associates have been brought up under more or less similar circumstances. The pattern of group life of the younger, sometimes of the older, in the more disorganized areas of the large metropolis is definitely negative if not also definitely antisocial. The gang, whether composed of teen-age youngsters or physically mature individuals, is a natural outgrowth of the multiple and complex set of forces which play upon the individual lives of each member composing the group.

If there are few or no opportunities to engage in wholesome activities, the normal, healthy human being, whatever his age, is likely to spend his or her energies in ways that are less than wholesome. In the case of the developing adolescent, his time must be consumed with activities that will give reasonable assurance that the desired type of social reactions will be built into him. If, on the other hand, during the entire development period of his life, negative, antisocial forces rather than the positive ones have been in the ascendancy, the individual probably will fail to appreciate the approved social patterns.

The history of boys' gangs and the record of delinquency left by them in some of our large cities is ample evidence of the fact that many parts of our great cities are no longer suited to the development of a wholesome childhood. In his studies of gang life and delinquency rates in Chicago, Clifford R. Shaw has found that certain areas have always been productive of a high rate of delinquency regardless of the nationality or race of the particular group living in these areas at any particular time. Whether native white, German, Scandinavian, Russian, Polish, Lithuanian, Italian, or Negro, each group that has successively composed the dominant part of certain of these areas from time to time has produced the same history of delinquency.

So great are the motivating forces of some of these areas that some students of the problem have concluded that the quality of the home life, whether the home is broken or not, makes relatively little difference in the type of social product that is developed. Such studies seem to reveal that a high rate of delinquency has always prevailed in such areas regardless of whether a small proportion or a large proportion of the homes have one or both parents living in the home. The ecological forces of the human community in such instances have, it would appear, produced a social setting out of which the human life that emerges will not have the type of behavior pattern or emotional responses that will enable it to function responsibly in society.

As the child in the cheap-rent districts of our metropolitan centers grows older, more and more of his waking hours are taken up with his group. Sooner or later this group assumes a solidarity through continuing association. Each youth comprising the group tends to have a function and a role to perform. As each strives to develop his own personality in the group, someone generally emerges as the leader. The others find their places through the process of association. In this way the gang life of the modern large city is evolved.

The gang is the product of that same artificiality of life in the large city that we have observed in the struggle to get a living. In the case of the gang, it is the attempt on the part of the youth, and adult also, for that matter, to find expression for normal, natural drives, for wholesome association, companionship, and a full life. When that opportunity is denied the individual because of the high degree of artificiality of his environment, the desire finds expression through the outlet of the gang with all of its warped ideas and ideals.

This relative denial of social expression in harmony with the realities of life encountered widely by city youth may perhaps best be illustrated in the relatively inadequate facilities provided in most large cities for play. Even though play is one of the most natural desires of the developing youth, what proportion of the youth in the cheap-rent areas of large cities has even a reasonable opportunity to give self-expression to this one natural urge?

Children and young people of the cheap-rent districts rarely ever have the opportunity to engage in the various types of games. For the most part, space is inadequate to permit even the simplest of games. Although baseball, for example, is almost universally the preferred game for boys, relatively few city youths get the opportunity to participate, except occasionally on a school playground. Denied the opportunity, the boy does the next most attractive thing and becomes a jaded fan—a follower of the game as it is played by professionals—by either reading the sports columns of the newspapers or trying to find ways and means of seeing the game in action, or both.

The boy in the smaller place, and especially in the open country, on the other hand, has abundant opportunity, if so inclined, to participate in the game. The fact that many players in the major leagues originally lived on farms is an indication of the fact that they had the opportunity to develop skill in the game at a comparatively early age.

It is in the rooming-house area that the greatest concentration of adult offenders is found. This is the major area of social and personal disorganization. Because of the close proximity of the area to the business district, rents are usually low. It is an area in transition from residential to business property. The large dwellings with rooms to let; the world of anonymous contacts and transitory relationships, the close proximity to the business district—all of these provide an excellent setting for the development of antisocial elements.

Ideally suited to the harboring of all types of criminals, the rooming-house area has become the center of the lawlessness in every form in all large centers of population. Gang life, for the young as well as the old, finds a natural breeding place here. Criminal gangs find the opportunity greatest for their activities. The red-light district and the centers of vice generally are found here. In prohibition days, the area furnished the locale for those engaged in the liquor traffic. Narcotic rings flourish because of the ready sale of their products to the mentally and socially maladjusted, whose numbers are usually large in such areas. The natural ecological forces which have produced these areas of disorganization in the first place operate to perpetuate the pattern of life that is dominant in the rooming-house districts. Waves of offenders may be arrested daily from the same address; repeated calls by the police, in fact, seem to effect no appreciable change for the better. On a broader scale, the attempts to make the area into a better and more wholesome place to live generally fail.

Admittedly, the rooming-house area of a large city is an extreme form of social blight, yet if we turn to some studies of crime and delinquency rates for the city as a whole and compare them with those prevailing for rural and smaller communities, the facts are none too reassuring. Let us consider, first, the rates of rural and of urban areas; second, the differential rates within the metropolitan community.

The urban population has a higher crime rate than the rural population. The difference, however, in the crime rates in small towns and in the open country is insignificant. Not only is the crime rate higher in urban than in rural communities, but also the rate tends to increase as cities increase in size.[4] In 1939, for example, the number of robberies increased from a rate of only 39.9 per 100,000 of the general population in cities of less than 10,000 population to 172.8 per 100,000 in cities with a population of over 250,000. In other words, the rate in the larger metropolitan communities was over four times the rate in the places under 10,000. Statistics showing the crime rates for other offenses follow closely the

[4] Sutherland, *Principles of Criminology*, page 122.

rates for robbery. From various studies with respect to crime rates for all types of offenses, including misdemeanors as well as felonies, the conclusion may very definitely be stated that not only is the over-all crime rate for cities greater than for rural areas, but that the rate tends to increase in harmony with the size of the urban area.

One distinguishing feature, however, of urban crime is the wide diversity in the types of offenses. In general, offenses in cities run primarily towards gainful offenses and offenses against the moral code. For that reason the proportion of misdemeanors to the total of all offenses in larger urban areas tends to run higher than the proportion of misdemeanors to the total of all offenses in the smaller cities. In 1923, for example, cities of from 25,000 to 100,000 population had a higher commitment rate to state and federal prisons and reformatories than the larger metropolitan areas. Since only those convicted of major offenses are committed to state and federal prisons, the inference may be drawn that the population of cities of small to medium size tend to commit more than their proportionate share of the serious crimes. According to the same study, the rate for the rural populations, which includes places of less than 2,500 persons, was not only lower than the rate for cities of from 25,000 to 100,000 population, but also distinctly lower than the rates for the larger metropolitan communities.

Other findings show that delinquency rates tend to be the highest near the center of the city and to decrease with the increase in the distance from the center of the business district. This fact tends to bring into their true perspective the disorganizing forces at work in the area of a city that is immediately outside or on the fringes of the business district. Formerly the dwellings in this area were owned by their occupants. When the community was smaller, the well-to-do generally lived here. As the city grew, these homes were too close in to the business houses to be considered desirable. As a consequence, their owners moved out to more desirable sections. In their places came persons wishing to take over the dwellings for the profits involved in renting rooms to detached single individuals. As a result of this tendency the area in transition from residence to business in most of our large cities has come to be inhabited by an anonymous population, for the most part composed of single persons who find employment in the business district. Since this area is predominately a world of young people, the actual number of children in these dwelling units is relatively smaller than in the homes adjoining where the working classes live. In terms of their numbers, the children in the rooming-house districts who get into difficulties with the law compose a higher averate rate of delinquency than is found in the zone of workingmen's homes. Whereas the concentration of delinquencies in the latter area is greater because of the presence of a much larger number of children, the rate of delinquency is higher in the rooming-house districts.

Most specialists in mental diseases agree that the incidence of mental disease has been gradually increasing in the United States for at least the past half-century. The period of most pronounced increase roughly parallels the period of rapid growth of cities.

The city has become the focus of a vast number of persons who crave excitement. These individuals find in the life of the large city a chance to satisfy various types of unsatisfied wishes. They are pulled to the vortex of city life much in the same manner as the light attracts the insects on a summer night. It is perhaps primarily for this reason that the city has tended to draw a certain type of restless, maladjusted, and neurotic persons. The smaller places, and especially the American countryside, would naturally fail to offer sufficient attraction to individuals of the type described. As a result they tend to migrate to the larger centers of population, where in turn they largely take up residence in rooming-house districts. It should be observed here that the element of the population in question is composed primarily of single, detached persons who are without family ties of any degree. These persons can be found congregated in large numbers in the rooming-house areas of any of the larger cities. It is in these areas, also, that the bulk of the major problems of city life tend to concentrate. Perhaps the relatively high rate of mental disorder commonly prevalent in such areas is one reason for their being centers of delinquency, vice, and crime.

One of the most outstanding studies of insanity rates within a city has been made in Chicago.[5] The study included the places of residence of 41,933 persons from Chicago who were committed to hospitals for treatment of mental diseases over a period of approximately fifteen years beginning in 1922. The rate ranged from 48 per 100,000 of the general population over fourteen years of age in the neighborhood with the lowest rate to 499. In other words, one neighborhood had ten times the rate of another within the same city. In general the rate varied more or less directly with the distance from the central business district, with the chief area of concentration centering in the rooming-house area.

While great variations in the rates of insanity have been found within the large metropolitan area, equally significant differences are found between densely populated states and more rural states. In 1939 for example, the number of patients in hospitals for the mentally diseased ranged from only 177.4 per 100,000 of the general population in New Mexico to 581.9 in New York.[6] The totally urban District of Columbia had at the same time the abnormally high rate of 996.9 per 100,000 population. In other words, one person in about 100 in the District of Columbia was a patient at that time in a hospital for treatment of mental disease.

[5] Ogburn and Nimkoff, *Sociology*, page 226.

[6] U. S. Census, 1939, *U. S. Summary of Vital Statistics*, Part II, page 1617.

The preceding figures measure the rate of resident patient population as of a given date. When considered in terms of a normal lifetime, what are the chances, in the more urban states, of an individual's passing at least some time in a hospital for mental disease? Studies made in New York and in Massachusetts showed that one person in every twenty of high school age in 1935 would likely be placed in a hospital for mental disease sometime during his lifetime.[7] The chances of developing serious mental disorder, however, sometime during his lifetime were one in ten.

[7] Ogburn and Nimkoff, *op. cit.*, page 214.

8. Economic Verities

S. C. McCONAHEY

GOVERNING the physical universe is a body of natural law which operates day and night over the ages. No less certain is the reality of fundamental cause and effect in human relationships. Social and economic conditions for any given period originate in underlying forces, some of them historical and remote, others more immediate and readily discernible. The first duty of intelligent men is to understand clearly and to interpret faithfully the forces creating control and consequence in this area just as surely as in the mechanism of the material world.

Lest this imply a mere mechanistic ordering of human affairs, however, we note that man himself creates his own social and economic environment, whether in wisdom or ignorance; and therefore by intelligent thinking and action may operate powerfully to influence the conditions that insure the highest social good.

In the interplay of social and economic forces the "gravitational pull and pressure" is forever towards ultimate balance or equilibrium, as has been brought out by Paul B. Sears in earlier pages. Yet today these forces have reached an extremely critical stage of strain or "pressure," first, because of the changed character of our national economy, but in a much more important sense because of artificial attempts that prevent the operation of natural law in the restoration of wholesome and equitable equilibriums.

During a period of less than 150 years, and accentuated over the past forty years, we have imposed upon the old, predominantly agricultural or land-use system of national sustenance a new and revolutionary type of economy as a consequence of machine-age, industrialized, mass-production developments. As recently as 1875, one-half of the population of the United States was still attached to the land or remained a part of the rural economy in small-town areas, the remaining one-half being dependent upon industry and urban occupation away from the land. Today fewer than 25 per cent of our people live on the land; and the simple fact that so large a proportion of our population is now detached from the self-sufficiency of land use and geared to the instabilities of a machine-age, industrialized economy poses the most vital single problem for this nation in time of peace. Military service and the peak production requirements of a highly mechanized war temporarily disposed of unemployment, but they also further dislocated the distribution of population and intensified the problems of peacetime adjustments, employment, public relief, and the participation of government.

The root source of most of our present-day economic, social, and political problems lies in the *unbalanced distribution of population as between rural and urban occupations.* Increasingly with the evolution of the machine age this unbalance has developed between the country and city, as illustrated by the chart sequence. We are now finally confronted with a condition that is unnecessary, illogical, and dangerous.

The inevitable consequences of this basic unbalance threaten our freedoms, including the freedom of individual enterprise and representative government itself, as heretofore relied upon for justice and the greatest production and distribution of real wealth that the world has ever known.

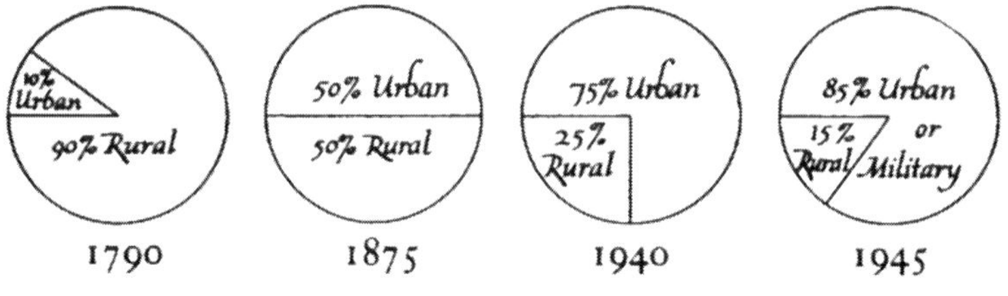

A familiar but highly superficial explanation of this shift from a rural to an urban economy is that mechanization has reduced the need for manpower in the country. This theory ignores, first, the fact that the same process of mechanization has also reduced the need for manpower in the city; but profoundly more significant, the fact that urban activities deal in large part with marginal, postponable goods and services, that is, luxuries, semi-necessities, and the specialized products of the arts, crafts, and sciences far removed from basic subsistence. These postponable goods and services of urban centers are the first to be sacrificed when depressions occur.

A much more accurate explanation of the trend we are discussing is found in the plausible allure of urban activity in terms of relatively high dollar income, shorter hours, less arduous physical work, bright lights, paved streets, freedom from isolation, overtime, vacations with pay, and, in general, an easier life with promise at the same time of a higher standard of living. But thousands of American families have discovered through bitter experience that these enticements are deceptive and that too often periodic unemployment destroys most of the apparent advantages of urban occupations and denies even basic subsistence. Obviously, the higher we build the urban ratio and economic dependence upon marginal and postponable goods and services, the more we increase instability and the perils of depression cycles.

The plain truth is that we have expected a progressively smaller rural population to spend more man hours of labor for less pay than urban dwellers in the business of providing goods, services, and basic sustenance for all of us. Those who ignore the potentialities of this critical unbalance between our rural and urban economies betray a strange mental illusion in the amiable assumption that the farmer will continue volume production indefinitely on the basis of a work week of sixty to eighty hours while his urban brother demands a constantly higher standard of living for half as many work hours contributed.

This is an extreme form of social parasitism, fraught with serious dangers to the individual and to the state.

We are concerned here chiefly with the effects of this unbalance in terms of the elements of basic subsistence for the family, that is, food, clothing, shelter. If the controlling economy brings such basic subsistence within reach of all who are willing to work, the social fabric will hold and resist indefinitely socialism, communism, or worse. At this vital point the capacity of a rural or land-use economy is practically unlimited because of the opportunity to provide for needs by family labor independent of dollar income. On the contrary, the economy of the city and industrial areas depends almost entirely upon an adequate flow of dollars, with disastrous results to individual and family life, the social structure, and the stability of government itself when this continuity is broken.

The theory that this unbalance between a rural or land-use economy versus an urban, industrialized, machine-age system must be maintained at all costs, we conceive to be a great fallacy in modern thinking which has led us into a wide range of government subsidies and related forms of artificial respiration. In truth, and of necessity, this program invades human freedoms and, still worse, substitutes the responsibility and judgments of government bureaucracies for the decision and responsibility of the individual for his own welfare. In brief, we arrive immediately at the beginning of a planned economy.

Citizens who lend support to these artificialities and disregard the unbalance which has grown upon us fall into two groups: those who believe that the unbalance can be maintained successfully in terms of "full" employment, through the operation of an "expanding economy," that is, a steady and continuous increase in the standard of living (production and consumption of goods and services); and, second, those who believe that government responsibility, planning, and control are essential in our complex modern society to insure the greatest good to the greatest number.

Regarding the view of the first group, the writer is sharply conscious, having had a lifetime of experience in manufacturing industry, of the importance of an ever-increasing volume of urban-produced high-standard-of-living goods and services. This importance not only relates to the maintenance of employment and a satisfactory national income but provides a formula for the widest possible distribution of real wealth. Nevertheless, while it is true, theoretically, that there is no predictable or calculable limit to consumption, the needs and desires of the people five years or a generation hence cannot be charted accurately in advance. Who can visualize the specific forms and direction of future inventions, or the psychological vagaries of the buying public, or the ability of customers to buy? Of course, our wants always run far ahead of purchasing power, but it is childish to contemplate an economy subject to such natural human instincts without an underlying system by which the exchange of goods and services is maintained soundly in relative equilibrium.

The reality is supply and demand; and the only safe, sound, and automatic control of constantly necessary adjustments in the exchange of these products of human effort is through the operation of a free enterprise system. In the very nature of the case, however, the principle of supply and demand imposes limitations upon both production and consumption.

In a word, an urban economy alone has definite limitations with respect to the volume of employment opportunity available at any given time. Moreover, it is subject to periodic instability which responds indifferently to cures applied at the industrial center rather than at the source. "Full employment" is possible, not in what has loosely been called an "expanding economy," referring, in common usage, to urban activities, but in a balanced economy—an economy which can neither forget nor forsake its rural base.

Regarding the point of view of the second group—those who believe that government responsibility, planning, and control are essential for the greatest good of the greatest number—the hardships of depression cycles make fertile soil for the propaganda that an all-wise and all-powerful state can and must supply plausible remedies by substituting for self-reliance and the free enterprise principle a planned economy promising security, if not happiness, from the cradle to the grave. Advocates of preliminary steps that lead to state socialism speak only of safety, security, and the responsibility of government to insure such guarantees to all, but ignore the loss of individual liberty that follows inevitably upon such transfer of responsibility for life and living. This is not, however, the philosophy of a free, courageous, and independent people; nor of that system of competitive economic struggle that has built this country from a series of small weak colonies to a great and powerful nation.

Those who are dedicated to planning would support the unbalance between our rural and urban systems of living by providing "purchasing power" through government debt and deficit spending and by other controls and directives. But purchasing power so created out of government subsidies is artificial and unsound, expanding the economy momentarily like a bubble, without substance or permanence.

I submit that the goal of economic independence must be the product of individual energy, will power, and work; and not economic dependence on government. Nevertheless, confusion and uncertainty have characterized recent decades because we fail to comprehend the cause and effect of economic forces. Economists have suggested that the social order suffers not from "economic maturity" but from immaturity in our economic thinking. Moral progress and the whole range of cultural values in human life, as well as the means of subsistence, depend, in final analysis, upon economic principles. Social discontent and political radicalism stem from serious weaknesses in our economic system. A basic weakness is the present extreme unbalance in the distribution of population as between rural and urban occupations and production, with periodic unemployment in dangerous volume as one inevitable result.

In order to visualize more specifically the situation in which we find ourselves as a nation, we define the two types of economy mentioned, as follows:

Agricultural, rural, or land-use economy comprises occupations, in whole or in part, on or closely related to the land; assures basic subsistence from the soil through the labor of the proprietor and his family and therefore a degree of confidence and security not otherwise possible; contemplates ownership of home and facilities as a rule, with the contentment, pride of possession, and independence inherent in such conditions of living; involves long hours, hard physical work; offers only a minimum of cash income as compared with many professions and other business activities; affords space and wholesome surroundings for bringing up children; no longer involves isolation from people and events; operates under a sound philosophy of keeping youth usefully busy as soon as physically able and old people employed as long as they can work; reduces the necessity for public relief to a minimum; and, in sum, largely offsets and corrects the weaknesses of the machine-age industrial economy which now dominates national life; develops social, economic, and political stability, and a high type of responsible citizenship, which is the strength of any nation and a guarantee of sound and liberal government of, by, and for the people.

Machine-age, mass-production, dollar-income (detached-from-the-land) economy, characterized by congestion in industrial and city areas, rented homes as a rule, dependence on daily dollar income and continuity of employment for adequate standard of living; yet, by and large, has resulted in an enormous distribution of wealth to all of the people; normally affords very limited facilities or none for using own labor to produce basic subsistence; is highly vulnerable to economic dislocations and periods of unemployment; is frequently subject to the demoralizing necessity for public relief; offers a deceptively high wage rate and limited hours of work as compared with rural occupations; encourages an artificial philosophy of life, including deferred employment for youth and early retirement for adults; in sum, is dangerously unstable, in large measure without the saving sense of a stake in and personal responsibility for the social order, subject to dislocations and incidental hardships creating political pressures and fertile ground for discontent and violence when inevitable depressions occur.

Successful representative government demands of a people courage, endurance, vision, intelligence, and action. Therefore, it is imperative that there be national understanding of conditions creating forces and trends. To know what is true and what is false is the first requirement for a sound economic system and for keeping men free. When truth is seen clearly in a free society, right remedies for wrong tendencies will follow.

Life under a rural or land-use economy is not without limitations and disadvantages, whereas the machine-age, industrial, mass-production economy in the United States has created the highest standard of living and the widest distribution of useful goods and services yet known. The problem then is to devise a program which retains the advantages inherent in each system. In essence we would combine the safety, stability, and cultural values of life rooted in the land with the capacity of the machine-age economy to produce goods and services in such quantity and at a cost to the consumer so low that the widest distribution of wealth will be achieved.

Now it is completely obvious that we cannot and should not attempt to reverse the course of events and return a majority of American families to dependence on commercial farming. While the machine-age economy, with its applications of power, has enabled the farmer to increase production in total volume and per unit of labor, it also has demanded a substantial capital investment and the training and ability to manage what is in fact a complicated and highly technical competitive business. On the other hand, for the great majority of families dependent upon industrial and urban occupation detached from the land, there is not merely a criminal waste of human energy and a dissipation of savings during inevitable periods of depression, but forces are released in the direction of government control and responsibility that if long continued will certainly end in a collective totalitarian state and the loss of individual freedoms which men have fought again and again to preserve.

The concrete suggestion that we present, therefore, is a compromise between life wholly dependent on land cultivation and complete dependence on the salary or daily dollar wage of industrial and urban employment. To provide what Elmer T. Peterson aptly calls the necessary "cushion" between the two types of economy here discussed, we submit the very simple and elementary "one foot on the land" concept for urban and industrial workers, but particularly applicable to and necessary for the lower income groups; that is, small plots of from one to five acres in cheap land zones bordering the great cities and crowded industrial areas, with low cost, cottage type, largely prefabricated housing, but including the modern labor-saving conveniences which the machine age has made possible. Good roads and cheap railroad and automotive transportation now make this program entirely practical, permitting continuance of industrial and urban employment so long as such employment is available. At the same time the family unit is in position to produce a substantial part of basic subsistence, not by the expenditure of dollar income, but with its own labor. Total investment contemplated per semirural family is in the range from $3,500 to $5,000, which is within the financial reach of a large proportion of industrial and urban workers. If, for example, one-half of our industrial families were so situated—with a vegetable garden, root crops, grapevines, small fruits, poultry, a pair of milk goats, and roses over the doorway—the periodic relief problem would be very greatly minimized.

The demand for socially demoralizing subsidies in the effort to maintain living standards and compensate for unemployment would lose the strong impetus it now has; and the social order would turn away from dependence upon government to independence and self-reliance. In promoting this wholesome and socially constructive program, education should lead in demonstrating possibilities to adults and youth alike; banks and other loan associations find opportunity to develop profitable financing; and business and industry discover the vision and leadership to encourage this adjustment to our times, thereby safeguarding and preserving the integrity and virility of a system in which all have a positive stake.

Let no shortsighted opposition to such a program from the commercial agricultural interests deter us on the fallacy that the market for farm products is foreshortened. When one section of national life suffers severely, as do industrial and urban workers in a depression, all suffer, including the rural resident. If a new and better adjustment between land use and industrial employment can limit depressions and accelerate the recovery period, all benefit. In fact, in a genuine free-enterprise system subject to necessary variations in public need and demand, interim periods of adjustment to realities and a sounder scale of values, are a natural, and even a desirable, part of the process—call these adjustments depressions if you will. The essential is that our family units have the will, courage, hardihood, and foresight to keep such temporary dislocations within bearable limits; and this a willingness to work and good management in land use can do.

Nor do we subscribe without qualification to the theory and ambition of able and sincere groups who appear to accept the dictum that private enterprise must furnish employment to all willing to work or the government will do so. Reference is made to the necessity for some "sixty million" peacetime jobs to replace the peak reached in a great war effort. We believe that neither the government nor private urban enterprise can carry through such a program and guarantee a nation of 135,000,000 people continuity of employment over the years. Such continuity is beyond the capacity of our machine-age economy, but not beyond the ability of private enterprise as a whole if we include land use, which, fortunately for all concerned, is capable of practically unlimited expansion as here defined. The implication has been that commercial enterprise—chiefly business and industry—should be responsible for "full" employment, and so far as we have observed there is little or no emphasis on the undeveloped possibilities of part-time land use. In order to arrive at a national economy adequately balanced and safeguarded during so-called unemployment periods, land use must play its part.

What are economic verities, as here envisaged?

First: All materials and resources vest in nature—the air, the sea, the earth. Man merely discovers these resources and by energy and ingenuity adds the cost of his thought and labor to new shapes and combinations and thus develops products and incidental services having value for use and in exchange at a profit. Such values we call wealth.

Second: Basic subsistence—that is, food, clothing, and

shelter—is derived primarily and most directly from the land. Since subsistence is basic and the first essential before we can proceed to higher standards of living and cultural progress, it is the most important and fundamental element in any economy.

Third: An economic system is faulty and dangerous which permits any large proportion of the people to become dependent for basic subsistence on a dollar income that, in the very nature of free competitive enterprise, must be variable from time to time with the volume of human needs and demands, such needs and demands being based on more or less marginal goods and services.

Fourth: The machine-age economy has tremendous values in the progress of civilization that should be safeguarded and preserved, but also an inherent instability that can threaten the very foundations of government.

Fifth: The strongest guarantee of the preservation of our present capacity for progress and a higher standard of living lies in a relationship or adjustment between rural land-use economy and a machine-age, industrial economy which will most completely retain the advantages and modify the weaknesses of each.

Sixth: The highest degree of economic safety and security for the average family unit may be secured in a direct occupancy of small land plots and the opportunity to produce a large part of basic subsistence with its own labor while taking advantage of industrial and urban employment when available. Such manner of life not only affords a substantial margin of economic security, particularly in terms of food and shelter, but tends to regain those wholesome values in rural living that establish a sound, clear-thinking, responsible social order, and the best possible assurance of the continuation of a free society in which government remains the servant and not the master of the people.

In presenting these conclusions there is of necessity over-simplification of many complex, modern problems; but the real purpose of this analysis will be served if attention is centered on the elementary concepts set forth and the urgent importance of realistic adjustment between the two types of economy now present in our national life which are both complementary and competitive.

9. *Government of the People*

H. C. NIXON

IT is revealing and instructive to glance at the impact of World War II upon the external and internal workings of the national state. These workings directly involve the fate or future of government of, for, and by the people.

One of the most significant developments in the Western world since 1500 has been the official and unofficial growth of national states, with their national cultures, their national economic systems, and their national administrative systems, all radiating largely from giant urban centers of finance and government. Man, in his outlook, has ceased to be a part of the world as he was in the Middle Ages. He has become a part of a nation, and his song becomes a national song. He also has come finally to the loss of many local loyalties, fashioning his ways of work and life under the influence of nation-wide corporations and organizations, a national government, and national tastes. He is under the dominant sway of one or both of two partly competing bureaucracies—the bureaucracy of business and the bureaucracy of government. The clerical forces of these two bureaucracies have so increased in the United States that they surpass in numbers one-third of the nation's farmers and farm workers. As for power, President Hoover's Committee on Social Trends considered pertinent the question whether business would rule government or government would rule business or labor and science would rule both.

Modern wars have tended to emphasize national power and the spirit of nationalism. However, the United Nations furnished during the war an unprecedented example of the co-ordination of national powers through interallied councils and conferences. This resulted in many nations functioning as one for war purposes. This unified action had the support of peoples, even before it was begun. National majorities wanted the Moscow, Cairo, Teheran, and Yalta conferences before they occurred. This international working together in wartime was significantly human rather than legalistic in origin. It reflects a more earthy spirit than the mixture of academic ethics and nationalistic diplomacy of the Paris Peace Conference, at which I spent the year of 1919 as an ivory-tower underling.

World War II, much more than World War I, forced the Western man to look at the world and his place in the world as in prenationalist times. The Axis Powers impelled democratic peoples to look beyond their national frontiers and to cultivate a sense of world community. Millions read Willkie's *One World.* A host of authors take the world for their parish, writing under such titles as *Total Peace*, *United We Stand*, *Chaos or Peace*, *Road to Peace and Freedom*, *Conditions of Peace*, and the like. At the same time, insistence grows that the conduct of foreign affairs be democratically modernized and liberated from visionless diplomats. That meaning underlies comments by Sumner Welles, Henry Wallace, and Wendell Willkie, not to mention seasoned journalists like Irving Brant and Walter Lippmann, or left-wing critics like I. F. Stone and Robert Bendiner.

International relations of both war and peace have come more into the concern of masses of people than ever before, thanks partly to the force of necessity, partly to popular top leadership, and partly to a democratic response by press and radio. If world unity is to be achieved, it is highly important that this folk concern and this folk interest be utilized, even with consequent imperfections in abstract planning for the world's future. In the democratic process, people are more important than plans. Successful operation under the United

Nations Charter must spring from the people, not merely from governments.

Modern world-warfare has inevitably expanded national administrative authority in the United States, as elsewhere. The ramifying use of the tax power, the commerce power, and other constitutional powers had already centralized great administrative authority in the national capital under the mixed demands of technology and economic maladjustment. Then came the use of the war powers, with administration to cover millions of persons in the armed forces and billions upon billions of dollars for producing and moving supplies for mechanized warfare.

This task was of such astronomical proportions that Congress soon found itself taking a greater hand in the general aspects of administration than ever before in the nation's history. In spite of partisan politics and legislative jealousy, much of this was distinctly constructive and helpful toward the home-front war effort. Few will deny that the Truman Committee of the Senate, for instance, saved the government many millions on the production front. With much disharmony and a traditional faith in checks and balances, the legislative and executive branches, nevertheless, co-ordinated their functions in positive performance. This is healthy progress, with electorates behind both of these great branches of government. The functioning together of President and Congress would improve if the partisan critics of each would remember that both are elected.

Those who would be pessimistic over our frictions in government should bear in mind that a democracy cannot smoothly and quickly shift gears from negative government to positive government and remain a democracy. In fact, friction is one of the essentials or unavoidable luxuries of democracy. Only an entrenched dictatorship can move easily at high speed without the consent or complaint of those being taken for a ride. Government today must be strong or perish, and the only way to achieve and maintain both strong and popular government is a hard way, requiring patience as well as wisdom.

The drama of the national effort in wartime tends to turn the popular mind away from the scene of local government, which is more important than glamorous. Former Governor Charles Edison, of New Jersey, has pointedly observed that he found many fellow citizens most ready and willing to rush to Washington for periods of government service but most reluctant to pitch into clean-up jobs at home. This suggests that local self-government might pass from the picture by default, even amid howls about centralized encroachment. It takes more than talk to make local government tick.

One of the most disturbing pictures on the whole landscape of local government is that of our large cities. This is true in spite of their reduction of the civic rottenness and inefficiency, which were once described by James Bryce in *The American Commonwealth* and later by Lincoln Steffens in *The Shame of the Cities*. Much of the present misfitness of city government stems from the misfitness of cities in technological, economic, and social trends, as brought out by other writers in this volume. Besides magnifying vulnerability to modern warfare, our giant urban centers are clashing with the census returns, and someone has aptly observed that it is impossible to quarrel effectively with the census returns.

Many cities grew only slightly or actually declined in population between 1930 and 1940, while decentralized metropolitan areas built up around them, both for residential and industrial purposes. Typical changes for much of the whole country are illustrated by a study by John A. Perkins, of the University of Michigan. Surveying Michigan's leading fifteen urban counties, he found county population gains between 1930 and 1940 ranging from .8 of 1 per cent to 40 per cent, while city changes in these counties ranged from a loss of 11 per cent to a gain of 13 per cent. Detroit's rate of gain was just half that of its county, and 30 per cent of the cities of 10,000 or more lost inhabitants, while only four of the 83 counties showed a decrease. Since 1940 decentralized war industries increased this jerry-built picture of disproportions. My own city of Nashville, Tennessee, is only part of an urban area with industries and clusters of residences lying just outside the corporate limits. This situation has entailed a messy mixture of makeshifts, shortcomings, and duplications in law enforcement, fire protection, water service, sewage and garbage disposal, public education, and public finance. It suggests a crying need for a rationing of government, so as to distribute both the burdens and the benefits more evenly between the rim and the center of our metropolitan districts.

There are ways of meeting these problems without adhering to any blueprint uniformity. One answer might be through metropolitan planning and zoning. Another might be through enlarging the corporate limits and jurisdiction of the central city. Still another might consist in co-ordinating or streamlining the governmental functions of city and county, particularly in providing essential services. There are many examples of success in this type of co-ordination, notably at Denver, Colorado. City and county have one health service at Louisville, Kentucky. City and county tax collections are combined at Anniston, Alabama. There are examples of co-operation between county-seat cities and county authorities in providing modernized farmers' markets. The TVA has facilitated or encouraged town-county co-operation, as in educational facilities.

If government is to facilitate the good life, it is necessary and proper that further steps be taken to remove the "baffling disharmonies" within our sprawling urban communities as well as those between urban centers and rural environs.

The conspicuous problems of urban and suburban government should not, however, blind us to the importance of what may be broadly and roughly called rural government, whether involving county, town, village, or township. It was largely with such local units that self-government had its real beginnings in America. In the face of the urban and suburban trends, over a third of our population is still on farms or in rural villages. Our democracy is largely of a rural heritage. So far in American history the rural scene has been our chief seedbed for statesmen. Study of the war and the world has not precluded interest in the hinterland and the rural locale. Aside from novels, the recent books on world trends can be matched by numerous recent titles on regional, rural and local life, such as *Rural America*, *Small Town South*, *Red Hills and Cotton*, *The Bayous of Louisiana*, and the *Rivers of America* series.

Rural government must change and is changing in order to survive and fulfill its function. The movement for complete consolidation of small or weak counties, which was launched about twenty years ago, has had only meager results. The scheme made a good showing on paper, but it clashed with local traditions and met political obstacles. But there has been no little progress in reorganizing or consolidating functions and services within and between counties. Several states permit two or more counties to make contractual arrangements for joint action in projects and services. In Louisiana, where rural parishes (counties) predominate, state legislation enables parishes to act jointly in various matters, including public utilities, public improvements, flood control, reclamation, recreation, and educational facilities. Georgia, Mississippi, Kentucky, and, to some extent, Virginia have made moves in this direction.

Many rural counties in recent years have modernized their systems of finance, accounting, and civil service. Lane Lancaster, a University of Nebraska authority on rural government, observes that there is less corruption in this field than in the market place. While there is still much room for improvement, it is time to discard the generalization that the county is "the dark continent of American politics."

Silent and wholesome progress has been made in streamlining county officialdom, significantly in California as of wartime. The governing body of my native Alabama county has been reduced to three, who are elected at large instead of by districts as formerly. Virginia has consolidated county law-enforcement activities. Clyde F. Snider notes that in 1942 "Petroleum County, Montana adopted the manager form of government by popular vote, thereby becoming the first small rural county in the United States to accept the manager plan."

Local government is on the upgrade in many ways. It is making changes comparable to the changes which occurred in state governments during and following World War I, conspicuously through the leadership of such governors as Frank O. Lowden, Alfred E. Smith, and Harry Flood Byrd. These local improvements may equal or surpass the state changes in significance, though they are less dramatic and less likely to lead to presidential ambitions.

County government, as well as other units of local government, has ceased to operate in virtual isolation, as once was largely the case. It is partly vitalized financially and controlled administratively from the state capital and in certain indirect ways from the national capital. It also operates partly as an agency for both of these higher levels of government. In other words, American government has developed into a functional federalism from top to bottom. That is true of the inner organic workings, if not of the legalistic framework. It means that local government, like local business or local farming, is no longer sole master of its own destiny. Small governments, small businesses, and small farmers are inevitably linked with national political and economic policy. It is important that democratic wisdom be applied at the center so as to avoid and avert national strait-jacket controls over local activities, from both Wall Street and either end of

Pennsylvania Avenue, as well as the large Washington building that houses the "United Mine Workers of America."

Wise exercise of power at the center is of vital concern to the national electorate, which is the total of all the local electorates in the American political system. In essence our national government, all our state governments taken together, and all our local governments taken together are responsible to the same tens of millions of voters and potential voters. This continental electorate, through chosen leaders and representatives, creates national, state, and local layers of government and distributes power among these layers. These distributive relations are always changing and always subject to criticism. Political criticism is essential in a dynamic democracy. It has a responsibility to serve truth and the state and should rise above the role of back-seat driving.

This electorate makes mistakes and chooses leaders who make mistakes. But it responds to strong leadership in times of crises, though it is resistant to dictatorship at all times. In fact, America leads the nations in the amount and complexity of constitutional and legislative checks and balances to hedge against dictatorial usurpation. It was a pertinent observation by Lord Charnwood, I believe, that the American people are the most legally minded, if not the most legalistically minded, in the world. That system of checks and balances partly explains why we have so many lawyers in and out of government. It partly explains why we seem to require a far-flung bureaucracy to get things done. We have a big and complicated set of brakes on the governmental machine, and it takes a big and complicated machine to move effectively with those brakes. We are struggling toward simplification of both brakes and power. But simplification of the complicated is not simple, except by a dictator's hammer.

It is erroneous and too simple to say that centralized government is dictatorial and that decentralized government is democratic. There are big and little democracies as well as big and little tyrannies in the world today. I have known of the domination of the governments of southern country towns and counties by managers of industrial establishments controlled by absentee owners. National democratic power can correct, has corrected, certain abuses of this type. The national government mitigated local evils in Louisiana in Huey Long's time, and a United States Supreme Court decision stopped discriminatory taxation of the anti-Long press. It was national power that wrecked the post-Long machine.

On the other hand, economic pressure groups have offered threats to the democratic process in the national government, particularly since the Civil War. The diversity of these pressures has saved the Republic. It has also seemed true in past times and recently that sincere reformers in places of power have been so impatient to attain democratic goals as to lack patience with the democratic process. Braintrusters and experts should realize that stateways cannot with success clash violently with folkways. Furthermore, there is no centralized monopoly of intelligence and good will.

The best security for democratic society in the American tradition lies in a diversification of government, operating on all fronts or levels of our functional federalism. This is aptly emphasized in the recent writings of exponents of democratic thought like Pendleton Herring and A. T. Mason, who observes that if freedom and democracy are to prevail, "the functions of government must be shared by many political subdivisions and functionaries, as well as by non-governmental bodies such as trade unions and co-operative associations." Basis for a dynamic diversification in government can be derived from the social philosophy of the late Justice Holmes, who insisted upon wide constitutional limits for national and state legislative venturing.

Diversification in government, with a mixture of centralization and decentralization, is called for by corresponding diversifications in American economic and cultural life. These three diversifications are healthy and are suitable to America's physical and human geography.

Good government, like good education, aims at the service and development of all the people through the participation of all. Progressive government, like progressive education, emphasizes process, as well as goals, as essential to human dignity and the good society. Such a progressive process requires local units of participation. All the light and all the thinking cannot come from great centers of great teachers or great officials.

The county is the most effective and the most nearly universal unit for the preservation of local institutions. In my own South and elsewhere the county is in many ways the last stand of tradition, and counties cover the country like the dew. The county seat has increasingly become the "grass-roots" center of governmental administration. County government is the general agency that makes direct contact with the governed. The county courthouse is the one place in town where country people can go without feeling that they have to buy something. Men and women from farms and hamlets now go there for types of information or advice which their forebears customarily sought at a local post-office, now extinct. The larger place has supplanted the smaller as the center of an organic "natural community."

There are signs that the county unit may come to the point of merely serving administratively to dish out government, ready-made government, to the citizen. This tendency can become unfortunate, even in the face of immediate gains in efficiencies and economies of a higher centralized administration. The economy should serve man, not man the economy. This criterion should prevail in public policy.

The county must have a greater function than just that of commission agent for the state and national governments.

It must continue to be a political and policy unit, a bedrock institution for registering the consent of the governed.

It is important that our growing state government and our growing national government leave wide ranges of choices to county government, choices in planning local projects and managing local affairs, not to mention the general functioning of an elected governing body. It is urgent that through the counties and their governments America preserve a degree of variety in its domestic institutions. If Thomas Jefferson were here today he might speak emphatically, not for states' rights, but for county rights.

These local rights that I should associate with a modern Jefferson would not so much be the rights to obstruct as they would be the rights to encourage neighbors to live together in good neighborhoods, to enjoy the fruits of productive labor, and to develop culturally under the "illimitable freedom of the human mind." The positive use of such rights minimizes the rootless elements in the population and increases immunity from fascism and other doctrines of oppression. A break-up of community life featured the shaping of Hitler's public career, with beginnings in a rooming house among uprooted fellows.

Between World War I and World War II this nation manifested two unfortunate civic neglects. In a nationalistic chasing of dollars and a nationalistic chasing of problems, we forgot the world community and the local community, each as an organic entity.

It is necessary that this nation look both beyond and within its borders for wholesome survival. As the national center yields something to the world community, it must leave something to the local community. Only in that way can America's thousands of neighborhoods and their "grass-roots" counties be brought into mutual sympathy with the world. The sense of world community and the sense of local community must go together. The world is not principles. It is people, and it takes all kinds of people, including neighborhood groups, to make a world. The world way will spring from the folkway. The song of the world will be a song of the soil. So

I sing of local unity,
Of larger local loyalty,
To free my hamlet from monotony
And tune it to the cosmic unity.

[Since sending in this piece I have for recreational reading gone through the two volumes of Tocqueville's *Democracy in America,* thanks to the attractive edition brought out in 1945 by Alfred A. Knopf, Inc., under the editorship of Phillips Bradley. I might have used this work as a bible in writing my paper and pointed up every paragraph with a Tocqueville quotation. Fresh and applicable today are his observations of a hundred years ago. He emphasized locally controlled institutions, whether of government, press, or other establishment, as furnishing both strength and safety to American democracy, as offsetting or forestalling the dangers of overcentralization or the tyranny of national majorities. Moreover, in the America which he traveled over, the provincial districts were not subject to the control of any great metropolis. He foresaw, however, a metropolitan abnormality for the United States.]

10. *To Clear the Dross*

LOUIS BROMFIELD

IT is a long-established belief of historians that vast concentrations of population in cities are one of the principal factors in the weakening and eventual decline of great nations. Behind the theory—which is in reality a good deal more than a theory—lie the many causal factors that have been discussed in earlier pages of this book. It is not for nothing that the great preponderance of the leaders of every modern nation have come from small communities or agricultural areas. Fortunately, the rural areas have had the men to give to the nation, and from the point of view of the city the steady flow of new blood from the country and from overseas, in the form of immigration and emigration has provided manpower without which it might not have held its own.

Into our own cities were poured, during the nineteenth century and the beginning of the twentieth, vast "herds" of cheap immigrant labor drawn from the economically and politically oppressed areas of all Europe. It must not be overlooked that the bulk of these were encouraged to come here with the express purpose of keeping down labor costs. That is where dross gathered in the melting pot. To clear it, we need an effective catalyst, one that will produce the social results necessary to stability and permanence.

For nearly half a century these unfortunate people found they had merely exchanged one form of oppression in the Old World for another in the New. They were really no better housed; indeed, all too often the tenements and shacks into which they were moved were worse than the buildings which had sheltered them in Europe. They were paid starvation wages, and they acquired the names of "Hunkies,"

"Dagoes," and "Polacks," among the native American population. They were crowded together in dismal areas where their only social life ranged from the church on Sunday to the poolroom and saloon during the rest of the week. Many of them never learned to speak any English, and for at least a generation or two no process of assimilation whatever took place in these ramshackle-built cities, where skyscrapers, handsome bank buildings, and museums existed side by side with slums as dreary as any to be found in Liverpool or Warsaw or Bombay or Calcutta. And presently, speaking their own language or dialect, reading their own native-language newspapers, they achieved a disrepute, which if not entirely undeserved, had been produced not out of the elements of the stock which they represented but much more out of the abominable conditions in which they were forced to live from birth to death. The Irishman, politically inclined by nature, got a name for corruption in politics, the Jew and the Italian for being the source of most of our gangsters, bootleggers, and black-market operators. The truth is that most of them, huddled together under abysmal living conditions, with permanent poverty and the periodical threat of starvation hanging over them, with only the street corner, the saloon, and the poolroom as a social background, never had a chance. A few individuals emerged as active, good citizens and leaders, but the greater number were never able to overcome the sharp handicap of their environment.

All this may appear to be a harsh statement of the situation, but it is, I believe, no more or less than truth. Responsibility for the situation must be placed largely upon our great disorganized, unplanned cities and the carelessness and greed which brought them into existence.

It was inevitable that one day the nation should have to pay for the curse of her ramshackle cities in terms of crime, of social unrest, of class bitterness, of bossism, of a decline in the standards of health, intelligence, and morality. I have never yet found that Emerson was wrong in his philosophy of compensation, and under that law we are paying now, and will continue to pay in the years to come, for the greed, the recklessness, and the economic distortion of our exploitation of the fabulous natural and human resources of America. The greed for cheap labor crowded more and more people into already congested areas; the greed of individuals and corporations brought inevitably into existence not only crime and ill health but the growth of violent political and economic ideas imported in essence from areas in Europe where they were born inevitably of both political and economic depression.

It is already apparent that the great development of industry during the past century and one-half has been made at a heavy cost to the human race—in overproduction in certain fields and at certain times, in faulty distribution, in living conditions, but most of all in the economic instability of larger and larger segments of our population. Outside of the areas with shipping facilities for world commerce, industry alone has been responsible for the growth of that monstrosity the modern city.

I think it can be said with justice that the only element of the population which has made any real and permanent gain through the concentration of industries into vast cities like Detroit or Pittsburgh is that minute segment of the population which owned the earth upon which cities such as these were built. And in any decentralization of cities that is the only element of our population which would not gain in every possible sense. Followers of Henry George would hold that most of these owners of real estate are not entitled to the vast fortunes which they have acquired, not through their own initiative, brains or enterprise, but simply through the fact that they happened to own the land or purchased it, speculating or profiting upon the results of the brains, initiative, and honest hard work of more worthy citizens. Consequently the single taxers would assert they have no right to special consideration or protection. In any process of decentralization this tiny element of our nation will be the only one to protest and their outcry is scarcely worth consideration in face of the immense gains to be achieved by a more reasonable distribution of population.

I have not overlooked the fact that the problems implicit in any sudden process of decentralization are not quite so simple as they might appear at first glance. They involve reduction in tax income to the city itself, reduction of income from the shares of certain real-estate corporations, and the dislocation of certain banking enterprises, but all of these are adjustable over a long period of time, and the rewards to the nation as a whole in terms of health, living conditions, citizenship, and individual economic security far outweigh them.

The idea of decentralization of cities is by no means new. Economists, sociologists, and philosophers have thought and written upon decentralization for more than thirty years in America, and in European countries for an even longer period; but in the past the discussion has been limited largely to the realm of theory rather than to that of practical application and fact. The plans, the purposes, and the results have been the concern largely of the theorists rather than of the *doers*, of the professor rather than the workingman, the industrialist, and the white-collar and professional citizen, who are the ones directly affected. Recently, however, the whole idea of decentralization has come forward upon a new basis, gaining its support from the practical rather than the theoretical side. That this movement has been spontaneous and over a wide field makes it the more important and the prospect of final achievement far more likely and immediate.

Henry Ford, himself partly responsible for the concentration of the vast automobile industry in Detroit, was also one of the first to understand the perils and disadvantages of huge industrial-urban concentrations. For years past he has been an active supporter of the "one foot on the soil" plan of industrial development and the experiments he has made with establishing factories in small towns and agricultural areas have been eminently successful on the score of economic stability, security, better health, and less social restlessness. Recently other industrialists have come to believe that in factories established in uncrowded areas, where better housing, living, and educational conditions and more individual economic security are possible, there are certain to be greater working efficiency, both in quantity and quality, and a good deal less trouble from strikes and labor disputes. (Not the least evil effect of cities is the tension provoked by the noise, the congestion, and the constant petty annoyances of daily life in overcrowded and unhealthy areas.)

It has long been Ford's idea that every workingman should have a house and five acres or more of land which he *owned*. Ford, like many other industrialist thinkers, sees such an establishment both as an economic anchor and as a cure for the migratory, hand-to-mouth, fly-by-night character of much of our industrial labor. The landholdings of certain Ford employees in rural areas are not limited to five acres. Many of them operate average-size farms successfully, an enterprise which today is possible with the eight-hour, five-day week and vastly improved agricultural machinery. Many of the workingmen look upon their agricultural activities not as work but as exercise and recreation.

From the point of view of economic security, the European workingman with his ancient tradition of "one foot on the soil" is in a far better situation than his more highly paid American brother. It is not only that in normal times his living costs are less and very often in exact ratio to his wages on a different level from that of the American workingman, but that he has that individual economic security which comes of owning land and a home, the kind of economic security which is and must always be the bulwark of democracy. In this country large segments of our industrial population, crowded into the great cities, or migrating from city to city or state to state in search of seasonal employment, have arrived at a point of ultimate economic insecurity in which the worker has no individual savings, owns nothing, and the moment he is laid off finds himself upon the public relief or the equivalent in some half-dole, half-make-work program. The evils of the installment plan, which in normal times keeps a high percentage of this nation constantly in debt, flourish best in the big metropolitan industrial areas.

Let us take the case of the fairly typical industrial worker in America, placing him in a situation which has become all too familiar to members of his group. Let us call the man Joe Smith.

Likely as not, and especially in boom times with high wages, he and his family live in a rented house and buy absolutely everything they consume. Much of this—furniture, automobiles, even clothing, is bought upon the installment plan or upon some form of credit. Nearly always Joe Smith is in debt for his wages at least a month, perhaps many months, ahead. Suppose that suddenly, perhaps for some minor reason, production in the factory where he works is cut, and he is laid off. The very next day he and his family are virtually in the street, their credit cut off, the furniture and automobile only partly paid for. The furniture and automobile can, by law, be taken back by the dealer because payments are uncompleted. A man and his family can live without an automobile. They *can* live in a wretched fashion without furniture, but they cannot live without shelter, without clothing, and without food. The articles partly paid for can be reclaimed by the dealer who does not want them and, if there is an economic decline in progress, can find no market for them, as secondhand commodities. Joe Smith loses the money he has paid on installments, and the dealer or manufacturer loses what has not been paid. The case of Joe Smith multiplied a few times begins a small depression or accelerates one which has already begun, cutting down markets and consequently production, thereby causing more men in the same precarious state as Joe to be laid off. Too many of these city-dwelling men represent the same problem as Joe Smith and contribute to the growing sum total; and presently what at first is only a small hole in the dike becomes a flood of economic disaster and depression.

But the story does not end there. Joe Smith and his fellow workers with their families, perhaps hundreds of thousands or a few millions of them, have to be taken care of, either by outright relief or by some expensive make-work program, and this burden can be paid for only in higher and higher taxes, further depressing enterprise, markets, and production, affecting not only industry but banks, agriculture, distribution service, invested money, and the whole of the economic structure.

Since the Civil War we have never had any real prosperity in America, but only a series of booms and depressions, each in its essence equally destructive, each beginning with the economic instability of the individual and pyramiding on the one hand into a speculative economy based on overexpanded credit and the installment plan (which produces only a fictitious and unstable prosperity) or on the other pyramiding into a colossal catastrophic depression.

No nation in history, whether a dictatorship or a democracy, is ever more prosperous or sound than the individual citizens who go to make up the whole. The growing financial and economic instability of citizens in America is centered not in the smaller towns or in the agricultural communities but in our great industrial cities where there is little or no tradition of economic solvency and where high costs of distribution, taxes, overhead, and so on raise living costs to a point where all too often individual wages are unable to cope with them.

All too often because of higher real-estate taxes, distribution costs, and general overhead, the dollar of the urban industrial worker is worth sixty or seventy cents as opposed to the one hundred cents' value of the dollar of the worker in the agricultural or small-town areas. Because of this, the urban industrial worker is constantly clamoring for higher wages that will permit him to keep abreast of the higher living costs of city areas. When he achieves an increase in wages, the increase is passed on to the consumer, raising the living costs over the entire country and bringing the big city industrial worker and his dollar exactly back to where they began. The vicious circle then begins all over again, with a steadily growing inflation which eventually can only end when the bubble bursts and plunges the whole country once more into vicious deflation, depression, and vast unemployment.

Take on the other hand the case of millions of industrial workers in a country such as France. I have known many of them and know their cases at first hand. There are in France fewer great industrial cities in proportion to the population than in America. The average Frenchman dislikes life in tenements and overcrowded areas. He is traditionally thrifty. He is by nature a landholder, an *owner* of what he and his family possess. Because the cities are smaller, it is possible for him, even in the largest ones, to live with "one foot on the soil," for transportation is easier and quicker into areas where the tax values of land are not prohibitive to his ownership. (One paralyzing factor in the concentration of workers—both industrial and white-collar workers—in the crowded areas of our vast cities lies in the fact that real-estate values for miles outside the urban core of the city are so high that the worker cannot escape into more decent areas without going so far that transportation and the element of time make an escape from overcrowded areas an impossibility.)

Let us call our French workman Jean Bosquet. He owns a house and a few acres of land within reasonable distance of his daily job. He actually *owns* the house and perhaps an automobile. If he has no car, he is one of the countless passengers on those omnibus trains which each evening feed out from a city like Paris thousands of industrial and white-collar workers bound for the remote suburbs or even purely agricultural areas. All of them, like Jean Bosquet, are going to the little piece of land which they own, and upon which their children are brought up. There one finds fresh air, elbow room, better educational facilities, and security. On each little piece of land there are chickens, ducks, a pig or two, a couple of goats, and perhaps even a cow. On it Bosquet grows all the vegetables and fruits, the eggs and milk, and at least some of the meat that his family consumes. There is a neighborhood cinema and a cafe here and there but not a poolroom or saloon on every corner.

When Bosquet is laid off at his factory, he can dig in and avoid the humiliation of public relief or make-work jobs. There is plenty of food, a roof over his head. He *owns* his home, his land, his automobile. Very likely he has a little cash laid by. He does not become, overnight, a public charge. He does not immediately contribute the weight of his personal disaster to the sum total of the nation's depression. He does not burden the economy of the nation with taxes to provide him and his family, miserably, with food and shelter. Most of all he does not lose his human dignity and self-respect and independence and fall into that perilous and humiliating state of mind where he expects the government or the rest of the community to support him.

Jean Bosquet and his kind are responsible for the fact that France has been able to survive disaster after disaster—enemy invasion, inflation, deflation, and international economic depressions far worse than anything we have ever known in this country. Jean Bosquet is always there waiting, economically secure as a tiny unit. Multiply these Jean Bosquets by hundreds of thousands and it will help our understanding of the toughness and resilience of France as a nation. Multiply the case of our Joe Smith, living in crowded cities in a constant economically precarious state even in time of prosperity, and it is possible to understand why our depressions grow increasingly sudden and disastrous and why there are so many demands for guaranteed employment, social security, doles, make-work projects, industrial insurance, and other measures, good and bad, which all too often destroy democracy and lead further and further into the philosophy of paternalism or support by government of its citizens.

I return to the point that the security, economic or otherwise, of any nation is only as great as that of its individual citizens. I make the further point that as individual security declines and national economic stability decreases there is only one solution and that is the gradual decline into complete paternalism and a condition in which all property and enterprise are owned by the state and administered by it. We have already reached the point of diminishing returns in taxation and as a nation live in a psychopathic terror of unemployment. We are already well along a path where more and more of our economic substance, much of the capital which should be employed to provide new enterprises and support established ones, is being absorbed into the process of feeding, clothing, and sheltering large segments of our population which are without individual security.

The great bulk of this economic insecurity is to be found in our great industrial cities; the greater the city, the more the insecurity. The proportion is much less in rural and small-town areas. The only way out, the only solution, under existing conditions, and the one popularly accepted is that we must have guaranteed yearly wages (which solve nothing of the fundamental problem of insecurity in congested areas), more social security, a "cradle to the grave" security plan, make-work projects, food tickets, worker's transportation pay from city to city, and countless other sound plans or nostrums, all of which lead deeper and deeper into bureaucratic and paternalistic government, with the gradual and consequent loss of the most precious of all human commodities, the freedom of the individual to work where and when he pleases, at what job he likes, and the right to build his own life as he sees fit.

Despite the fact that by increasing our economically insecure urban population, both industrial and white-collar, we are constantly weakening our social and economic structure and undermining democracy itself, we are still treated to the spectacle of great cities competing with each other to bring in new industries and fresh thousands of industrial employees for the benefit in the long run only of the small element which holds title to the land on which the monstrous cities are built. Darwin, if he were alive today, would, I think, find a parallel between such a policy and the blind and foolish behavior of the lemmings of Norway which periodically migrate in millions to the sea where they swim out and out until, at last exhausted, they cease to exist.

Much bitter controversy has arisen over economic conditions in the nation during the past two decades and much bitter controversy over the economic and social measures taken to correct or ameliorate these conditions, but too much of it has been on a political level, ascribing an impetus to the political motives of individuals, groups, and political parties which are in fact merely the symptoms or manifestations of a situation which is fundamentally economic and partly social in its origin, arising out of the increasing economic instability of our industrial and consequently our urban population. The fundamental condition, out of which arises our psychopathic fear, as a nation, of insecurity and unemployment, gives rise to all the controversial measures, good and bad, which have been undertaken to meet and correct it. We should have had the same or similar measures under any political party or administration because the necessity for such emergency measures was at the time imperative. It was the undeniable emergency of providing food and shelter for human beings.

There were two ways of meeting and solving the economic insecurity of the individual and consequently that of the nation—one by providing greater security for the individual through his own efforts, opportunity, and initiative (the democratic fashion), the other to provide security for the individual through a government planned and executed economy. Of the two the second is, if not the happier, by far the easier, the quicker, and politically the more flashy and profitable, and it was that course which aroused resentment and rebellion among a large segment of our population, accustomed to the blessings of free enterprise and liberty associated with happier days when the whole country, incredibly rich in natural resources and industrially underdeveloped, was not yet face to face with the problem of a growing individual and national insecurity.

I return to the point that our national insecurity is largely to be identified with our great cities and the industrial population which inhabits them, and that the cure of the fundamental sickness in terms of democracy lies partly at least in the decentralization of these great, industrial concentrations of population and human enterprise and the dispersal of both populations and factories into rural areas and small urban developments, if we are not to go all the way along the road to paternalism and complete government ownership and control.

Many influences are working spontaneously and without much human guidance or purpose toward this decentralization and dispersal. As has been pointed out, both industry and organized labor are becoming aware of the evils of great industrial concentrations as regards income, living costs, efficiency, health, labor unrest, and individual and collective economic security. The Ford-Ferguson agricultural machinery corporation in Detroit, which has plans for a great expansion in the immediate future, will manufacture only its tractor at the Dearborn-Detroit plant. All the 150 agricultural implements which are a part of "the Ferguson System" will be manufactured in 150 small towns scattered over the United States. The rubber industry, crowded senselessly into a single city, has long had plans for decentralization under consideration.

Outside the high-value areas of all our middle-sized and small industrial cities there is everywhere in America, a rapidly growing band of small houses and small landholdings, like those bands which surround Paris and other French cities. These represent the spontaneous search of individual workers, both industrial and white-collar, for a security which they found tragically absent in the great depression of the thirties. Much agricultural land changed hands during the war, but the basis of purchase and sale has not been either speculative or expansive as in the last war, but represents the search for the security to be found by the "one foot on the soil" philosophy. Many family trust funds and some corporations have placed a part of their liquid capital in agricultural land in the knowledge that in time of extreme inflation, agricultural land represents almost the only stable investment and security for capital.

The figures of the latest census show a decline in population of many middle-sized industrial towns and even some large cities. Very often those figures are misleading in that they do not represent an actual loss through migration of population into another metropolitan-urban area but only a shift of a part of the existing population into the areas *outside* the existing city limits, into that growing band of small semi-agricultural holdings which surrounds more and more our large cities.

The growth and development of modern farm machinery has made it more and more possible for industrial and white-collar people to operate profitably—under a five-day, forty-hour-week work schedule—five to ten acre tracts and even fair-sized farms. With such projects go not only material security, but better diet, health, morality, and very often better educational facilities.

The development of the Tennessee Valley Authority has shown the way toward one important kind of decentralization, by taking over an area larger than the British Isles and developing it simultaneously along *both* agricultural and industrial lines, under a plan by which enterprises in both fields exist side by side without the disadvantages that exist in our crowded industrial areas. The plan has taken into consideration the fact that a sound industrial wage income can guarantee a small agricultural project and that a small agricultural project can guarantee security for a man and his family even under the most severe of economic depressions and that the two enterprises go hand in hand and together spell individual liberty, dignity, security, and independence.

Anyone in contact with the armed forces is aware of the great impulse among the men towards setting up small businesses of their own or acquiring small farms after the war. The impulse of countless soldiers is to re-establish themselves, after the war, not in the great industrial cities upon an economic basis of wages alone but in smaller towns and in agricultural areas where they will have both freedom of action as well as freedom from fear of depressions. At the same time, some careless thinkers have seen the wholesale re-establishment of returning veterans on the land as at least a partial solution to the problems of readjustment. This is a dangerous theory and if translated into a policy can only bring about great confusion, a growing bureaucracy, and countless personal tragedies and failures and much bitter disillusionment.

Agriculture is a difficult art and profession and business. A successful farmer must combat the economic vagaries of the market and the natural vagaries of the weather. He must know more about more things than any other member of society. He must be a mechanic, a botanist, a horticulturist, a veterinarian, a chemist, and many other things; and, most of all, he must have a natural inclination towards his occupation. The idea of placing great numbers of our returning soldiers or transferring elements of our urban population wholesale on to the land without any other resources is both impractical and filled with ominous possibilities. However, the idea of helping the returned soldier or the dissatisfied city dweller to establish himself and his family upon a small holding which can be worked in conjunction with a small business or an industrial job is a sound one. The one activity coordinates and guarantees the security of the other and together they provide him with economic security and a genuine stake in the economy of the nation. It is, however, a plan which is impractical for those who work in great industrial areas which involve transportation and time to go great distances to and from work.

A Missouri Valley Authority is now under consideration on a scale even larger than that of the Tennessee Valley. It opens up the prospect of a whole new decentralized industrial-agricultural area with towns established where there is abundant hydroelectric power with available agricultural land and the possibility of small holdings near at hand, of a whole region in which a new America can be developed to the mutual advantage of industry, labor and agriculture, free from the evils which are inevitably concentrated in vast industrial cities. It is possible and indeed probable that the TVA and the proposed MVA are showing the way towards a New America in which individual security will be a fact based upon economic and social common sense, rather than upon hasty, emergency measures which in the end do not cure but only aggravate through taxes, doles and relief, an economic situation which is fundamentally unsound.

The whole impulse towards "one foot on the soil," towards decentralization of vast industrial cities is a spontaneous one. It may well point the way to a sound solution of many of our social and economic problems. It is not the whole solution to our problems, but it contributes much toward the individual security upon which the stability of any nation and indeed of democracy itself is founded. Great cities inevitably produce populations which in the end can only be appeased by the "bread and circuses" perilous to the security and welfare of any nation. It is, I think, time to consider doing away with them, simply on a basis of common sense and social and economic security.

Long ago, even before Karl Marx saw the inevitable connection between dispossessed industrial populations dependent solely upon wages and periodic employment and the communist state, Thomas Jefferson understood and stated the connection between sound democracy and populations which owned something and so possessed a share in the nation's welfare and indeed in freedom itself. It seems to me that we are at the crossroads. Either we drift on and on into the depressing condition of a corporate state, or we act to establish a redistribution of economic values and continue as a democracy in which the rewards are free enterprise, independence, human dignity, and freedom.

11. *A Farm Reporter Looks Ahead*

LADD HAYSTEAD

WE like to think of American businessmen as the greatest selling group in existence. Most of us believe that American salesmen have developed all possible markets, searched out the far corners of the world in their hunt for customers, sold refrigerators to Eskimos and umbrellas to South African chiefs. The charge is often heard that they have oversold the home market, egging people on to buy things they can't afford and would be better off without. That picture isn't quite correct. American salesmen have failed to develop fully the basic domestic opportunity, namely the farm market.

Some sellers have looked at the national farm income, the average income per farm, and without further thought have dismissed the market as not worth while. Not enough buying power there, the judgment has been. This ignored the fact that a farmer has no charge against his income for shelter, often none for fuel, and better than a billion a year coming to him in the form of farm-raised foodstuffs which do not appear in his cash take. Add to this something over two billion dollars annually in cash returns from investments, boarders, roadside stands, off-the-farm earned income, and so on. On the other hand the farmer pays less for recreation than does a city man, for the farmer makes much of his own. Taxis, subways, and other transportation charges eat up part of the city man's cash income. Hence, the farmer often has much more spendable income than a city man with several times the farmer's annual take.

If we examine the extreme urban example, New York City, we find that better than half of the average worker's earnings goes for shelter and food alone. Another bit goes for city sales tax. A considerable amount for clothing, recreation, and medical attention leaves nothing or almost nothing for many of the products industry thinks the farmer cannot afford. When average farm income is less than a thousand dollars, it seems the farmer is practically a pauper. Yet the New Yorker with $2,400 to $3,000 will actually have practically no spendable cash after paying for shelter, food, clothing and recreation. In smaller communities this admittedly is not true, but in those communities incomes of better than $2,400 are far more rare than they are in the big city where everyone admits $50 a week is just about enough to keep a stenographer in necessities, including nylons.

As sellers did not realize that the farmer had relatively higher buying power on definitely smaller cash, industry designed everything for the city market and let the farmer make his own adaptation if he would. The different conditions surrounding farm life were overlooked. Design for the urbanite, ignore the ruralite, was the rule. A case at hand is the passenger car, strictly a city vehicle. It has no place in it for the few cases of eggs or several cans of milk which many a farmer carries to town every time he takes the car out. There is no place for the feed, insecticides, dips, hardware, and other items he brings home. Presently he loads the stuff in the back, tears up the upholstery, cramps the passenger-carrying capacity, and philosophically wonders why Detroit doesn't recognize that his automobile should have a dual function.

Detroit might come right back and say that the farmer is the biggest buyer of used cars, that the designers have to look to the first buyer, who is a city man. But with 38 per cent of all passenger cars registered in rural areas, it might be a profitable idea to see if this area could not be made into a new-car market by offering a farm-designed car. There is one hitch in the idea from industry's angle, however, for two old cars must be sold to make room for one new one and the farm is the No. 1 place to get rid of the secondhand car. This doesn't please some of the fighters in the co-operative movement. They say that if industry won't design and produce a dual-purpose car, then the co-operatives will take up the job. But at least one farm-equipment manufacturer now has an experimental model under way at a factory in Illinois.

The automobile is only one item. There are hundreds, probably thousands, of others which could find a much larger market on the farm if industry would study the peculiar needs of farmers and design items to fill those needs more expertly than is done at present. In some cases it means cutting out frills to produce a more inexpensive product. In others it means a whole new design. It has to be remembered that while the farmer can be sold much more than has been the case up to date, on the other hand he is less likely to buy the unnecessary than the urbanite whose buying decision may be more predicated on "keeping up with the Joneses" than it is on necessity or personal satisfaction.

If industry has been asleep at the switch in the past, it is waking up rapidly now. A number of firms used to have an agricultural engineer on the staff. For the most part, he was the fellow who had an office under the stairs, was rarely consulted, poorly paid, and no one was quite sure why he was around the place. In other cases concerns that did a really important farm business did not know it, or if they did, gave it scant heed. They had no agricultural engineer at all. There was no one to tell them that a goodly portion of their sales to urban jobbers actually went into farmers' hands as the ultimate consumers.

But all that is changed. Now the agricultural engineer has been moved out from under the stairs and given a paneled office right up among the big department heads. Anxious planning boards are listening to him, asking questions, and getting into gear to tackle the farm market in a peacetime world. Demand for trained agricultural engineers is greater than the supply for the first time in years. As a result, the newly discovered farmer is about to feel the greatest attack on his pocketbook and appeal to his desires in all his history.

This time he will not be given city things but will find many and many a time-saving, labor-saving, comfort-making gadget engineered directly for his requirements. Planners are busy not just dreaming up Jules Verne-ish words, but actually getting things down on drawing boards. They are building mock-ups, experimenting with models, trying out in the field "ghost" editions of what are expected to be tomorrow's standards.

The field is so large, the plans so many, some breakdown must be made to cover the topic adequately and in an organized manner. Therefore, the topic is split between the two major facets of farming: (1) as a way of life, and (2) as a means of production. The philosophic and spiritual angles of the components of this split are treated expertly elsewhere. Here, we stick strictly to the materialistic.

What's Ahead for Farming as a Way of Life

The natural starting place is the farm home. It has been falling to pieces for over twenty years, or since the great farm depression started in 1921. Money put into maintenance and repair has been, for the most part, less than annual depreciation. Paint has been lacking. New sills, new doors, windows, hardware, siding, roofing, and—very important—new plumbing have not been given to existing structures. At the same time, many a structure has gone entirely by fire, windstorm, abandonment, or simply old age and lack of care. Today there exists an enormous market for building supplies estimated variously from a half-billion to two billion dollars per year for the next ten years.

At the present, over 40 per cent of all farms are electrified, but only 12 per cent have inside plumbing. The results of surveys made by utility systems, REA, and other sampling agencies show that by far the majority of farm women want inside plumbing more than anything else. And with farm income hitting ever new highs, with loans and notes being paid off at a great rate, with total mortgage debt descending and rural banks bulging with demand deposits, the only reason there won't be an enormous boom in plumbing supplies is sheer lack of salesmanship on the part of that industry.

Electrification will expand, not only to more farmsteads (REA estimates saturation at better than 85 per cent and thinks it may not take many years to reach that point) but on the individual farm where the power is already in. Usually the electricity is wired into just one room in the house and also into the barn where it can operate milking machines, water systems, feed grinders, and the like. But with a loosening up of materials and the providing of trained manpower to make the installations, plus salesmen to argue the benefits of electrified kitchens, radios, deep-freezers, and lights in every room, farm homes are going to be on a par with city buildings.

In this regard, it should be noted that a wholly electrified farm home does much more for the family than an electrified city home where the service trades perform so many chores. In the city, electricity lights the home, runs the refrigerator, the radio, maybe a few table appliances, and a vacuum sweeper. In the country, on the other hand, electricity provides running water as well as all the above, runs numerous labor-saving kitchen machines, the laundry, the deep-freezer, dehydrator, the churn, the water heater, the stove, roaster, bathroom heater, and possibly in the future an inexpensive air conditioner.

In the home where this is written, plug-in strips run around the kitchen walls just above the metal cabinets whose tops form the working surfaces and below the wall cabinets where everything is stored out of sight, away from flies and dust. The two plug-in strips provide outlets every eighteen inches. One carries a much higher amperage than the other to provide for the heavy loads heat appliances put on the line. Without moving a step, the kitchen worker can reach four or five of these outlets to plug in whatever one is wanted of a dozen or more electric appliances which make work in that kitchen a joy.

Overhead, fluorescent lights give a shadowless daylight after dark or on stormy days. An endless supply of steaming water is available at all times from the neat white-enameled water heater which fits right in as a unit of the metal floor cabinets. The electric stove and roaster are automatic. The green beans are sliced by an electric machine, the peas shelled, the meat ground, salad dressing whipped, potatoes mashed, foods prepared for the freezer or canning, all by machine. It is a fact, proved many times over, that a meal for a dozen can be prepared with less trouble in that kitchen than a meal for two in a New York apartment.

Each room has its radio so that various members of the family can listen to their favorite programs without interfering with each other. A house telephone system saves many steps as people can ring each other from basement to second floor, or from one end of the house to the other. In all workrooms, fluorescent light gives its eye-saving, nerve-saving benefits. The entire sprawling house is run by electricity from basement shop to upstairs study. And the cost averages about seven dollars per month.

That house is not a house of the future but was striving hard to keep up with progress when the war broke out. No doubt, if further changes and additions are not made in the next few years, it will be obsolete in a short time. Certainly, these helps to better farm living cost too much—more than the average prewar farmer could afford. But it is reasonable to assume from plans now known to be working out that a similar installation will tomorrow cost only a fraction of that of the sample presented. The amount of savings in labor, food storage, and preparation, plus the increase in living values, will make a matter of less than ten dollars a month, an amount many millions of farmers can and will afford.

Among the new gadgets which may be with us within a short time is the radio-telephone. It is not generally known that rural phones have been declining in numbers for some years. Anyone who lives on a country party line can testify to several reasons for finding the telephone more of a nuisance than a help. But the research men have a new idea in the works which sounds as if it would remove all difficulties. This idea is to carry the telephone waves on the power lines. A contraption which resembles a tiny sending and receiving outfit will be installed in the house. With this assembly, complete privacy will be the user's benefit, plus the fact that anywhere the power lines run, telephone service can be available. With 85 per cent saturation of farm market for power, that means that almost all rural homes can have the instrument most city people think indispensable.

Other developments are a small flour mill to make flour right on the premises. A new kind of heatless cooking is promised. This operates much like the short-wave diathermy machines. It means the end to sweaty days over the stove in summertime, while it speeds up cooking immensely and retains all minerals and vitamins. Still another device gets away from either the furnace or the base-burner. By the use of electronics, heat is shot at the bodies of people while room temperature stays away below what is now considered comfortable. The installation is simply two invisible poles set in the walls of the room. It is claimed that no one will need covers on the bed at night when the windows are open, yet complete comfort will be the lucky householder's situation if he has the new type of heating equipment. Again, cheap rural electric rates suggest this gadget as particularly adaptable to farms.

Farm homes need a tremendous amount of new furniture. New materials will help to bring these comforts and necessities to many more farms than in the past. No doubt design will change materially. The creators will recognize that moisture changes in the atmosphere are often very much greater in the country than in the city. Colors are inclined to fade faster because more direct sunlight is admitted to farm-home rooms. Clothing fresh from barn or fields tends to stain or dirty conventional materials more in the country than in the city. Floor coverings should not have a heavy nap, should not be dust or mud catchers. The velour and plushy drapes so dear to the city woman are not suitable for the farm. These and many more things are being considered by the modern, farm-aware designers as they visualize the farm home of the future.

Even the landscape gardener finds himself with new customers. When farm homes get running water and are electrified, farm folk will want to take the next step and beautify their surroundings. They did not do without lawns and flower gardens because they had no esthetic appreciation of them. They didn't suffer searing summer winds and bitter winter gales, when they might have had tree protection, because they did not know any better. To landscape a place, you need water and plenty of it. You need time to work among the shrubs and trees. In the past, there was neither water nor time. But with electricity, the farm home can have automatic watering devices. With rotary tillage implements, which are now recognized as excellent supplements to heavy power, a farmstead can be landscaped and maintained with less trouble than a place in the city, to say nothing of the advantages of a plentiful supply of manure which most farms can have.

The farm home has always been by necessity more or less an educational center. In the future it will be vastly more so. Lessons will come to shut-in youngsters from a central educational broadcasting studio. When blizzards block the roads, the day's educational grist will come right out of the loudspeaker or from the television set in the living room. Great plays, operas, concerts, and lectures will be made available to new millions.

Farm practices themselves will be more adequately demonstrated than ever before right in the farm home. Where possibly only a few could journey to town, the county seat, or the state land grant college to learn how to combat a new pest, in the future that pest will no more than make its appearance before the farmer will learn how to handle the situation on the television screen in the dining room while eating his lunch. If someone devises a new kind of tillage—and it's by no means certain that we have the best of all possible methods now—that new system can be demonstrated to millions in their own homes.

Whether or not the helicopter will become a part of farm life, as some have rushed into print to predict, it is too early to say. Certainly, even before the war, the bigger ranchers and farmers were pointing the way for others by their use of private planes. An air flivver or a helicopter may indeed be a part of the farm world of the future. If so, the last bit of isolationism (not in the political sense) will vanish from rural life. The motorcar did much to break down the old pattern. The airplane may complete the job.

But while the more conservative are inclined to go slow in prophesying an aero-boom comparable to the motorcar boom that followed the last war, it is a very good probability that a master highway system with vastly improved feeder roads will come into being. This in itself would just about bring the farm as near the urban centers as the airplane. When a farmer can swing off a feeder road, macadamized or better, on to a cloverleaf, and thence to a great express highway with no crossroads or other interference, he will be a hundred miles away in only a couple of hours. Except in the range country in the Far West, this is sufficient to put most farmers into a big urban center with requisite speed and dispatch. Of course, most planners think mid-size cities will become more metropolitan in their offerings, though smaller in size and greater in number in the future as decentralization of the huge places such as New York continues the observable trend. Thus, it is at least possible that nobody, including farmers, will care to travel any great distance, for there will be little need of it.

Certainly if vertical integration of farm industry proceeds and if rural industry grows as some think it is bound to do, there will be decreasing need of distant travel. In its place, the average farm community will become more self-sufficient. Feeds will not be mixed a long way off, but at home. Likewise, fertilizer may become a local fabricated product.

Some expect to see more and more assembly plants scattered all over the country close to the buyers rather than close to raw materials. Subcontracting is here to stay, think others. Instead of having all parts of various necessary machines made up in a center such as Detroit, it is possible that farmers will become factory workers during the winter, making various subassemblies themselves in small community shops. These in turn will be gathered at final assembly plants in larger communities, but much closer to the ultimate user than at present.

Some of the consumer co-operative thinkers are probing into this idea with more than just theoretical planning. They have the money. They have proved their point after a century of many defeats and few successes. They know they have a new power, and they are beginning to use it. It may be that at least some of the milking machines needed in a dairy region will be made right in that community by some of the people who are going to buy the product. It may be that poultry equipment will be made in the densest poultry areas such as Delmarva, Arkansas, Nebraska-Iowa,

Missouri, and New England. Summertime cowboys in the Mountain States may be wintertime builders of windmills, currying and dipping machines, tools for dehorning, castrating, and branding, and all the other paraphernalia of their trade. If so, community life is going to change. We'll see fewer large cities, more self-sufficient small towns and a more equitable distribution of trade, industry, and transportation.

The nation as a whole will then, of course, present an entirely different pattern than it has in the past. The long-time coagulation of indigestible quantities of people into relatively small areas will reverse. Little bankers will do more local financing, while they spread their risks and tap bigger quantities of capital by refinancing and discounting through the existing huge concerns. It is not so likely that bigness in all things will become historical as it is that we'll see more bankers' banks, more insurance company insurers, more industrial industries.

Some ideas of what may well be in store for rural communities is given in the *Fortune* Farm Column (October, 1944), republished by permission here.

Rural Industry: A Solution?[1]

The dreary lessons of the thirties are making themselves felt in the labor market as war workers tend to leave temporary high-paying jobs to look for permanency even at less cash return. Many soldiers have said that their greatest post-war worry is unemployment; labor leaders have dourly predicted a million unemployed by January.[2] Leaders in business, industry, agriculture, and community planning have been digging into odd occupational corners to find jobs for the greatest labor force in our history. One possibility has

[1] Copyright, *Time*, Inc., 1944.

[2] Forecast for January, 1946, a figure which has in fact been exceeded.

been only casually explored but deserves a great deal more attention: rural industry.

Many things can be done on the farm that are not done there now. True Morse, president of Doane Agricultural Service, St. Louis, Missouri, has suggested that farmers carry out the first steps in processing farm products. ("Buildings on the Farm," *Fortune*, May, 1944.) D. A. Milligan of Harry Ferguson, Inc., has proposed that farmers work part time on the land and part time in near-by factories. (The Farm Column, September, 1944.) Professor F. E. Price of Oregon State College believes farmers and their families should process flax for fibre and set up community plants for vegetables and fruits. REA sees possibilities of employing farmers by the hundreds of thousands in small, electrified industries.

Some rural industries would cost very little to set up. Existing farm machinery can be used for processing food, reclaiming waste products, or preparing raw materials for finished work elsewhere. For instance, straw could be run through the silage chopper preparatory to strawboard manufacture in a community mill. It would not be costly to clean, blanch, grade, and quick-freeze foodstuffs, or even to precook whole meals for door-to-door delivery in near-by towns.

On the other hand, small community soybean mills would cost about $50,000 to set up, whether chemical or mechanical methods of separating the oil and the meal are used. Yet they can amortize their costs in a very few years. In effect they let the neighborhood sell the oil and keep the meal, thus saving transportation costs. In the soybean belt, there are hundreds of communities where bean production is high enough to justify a mill in each township. While they do not offer high employment possibilities themselves, such mills would contribute materially to the community's economic health.

The use of farm products is not the sole interest of promoters of rural industry. For example, Cherokee, Oklahoma, recently became the home of a brassiere maker. A concern in Hollywood couldn't get labor. The manufacturer picked up his plant, leased a building in Cherokee, and brought a thousand-dollar weekly pay roll to the farming community. Since then, the same company has set up ten more units, each small enough to use only local labor.

Community industry in Mississippi got a boost in the late thirties by a plan to "Balance Agriculture With Industry." Farmers were unable to sell their produce profitably, and small towns were disintegrating under the joint pressure of low farm prices and unemployment. A state commission was set up, public funds were made available for building low-rent factories, and twenty industries were brought to the small towns of the state. Some inevitably died a-borning, and when the war came on, some were closed down for lack of labor and materials; but twelve remain. They manufacture hosiery, rubber tires, woolen goods, shirts, chenille products, plywood, and steel ships. According to a study published by the Federal Reserve Bank of Atlanta, the twelve B.A.W.I. plants in the middle of 1943 had 14 per cent of the employment and 24 per cent of the pay rolls of all Mississippi's manufacturing. The study says further that the plan convinced Mississippians that "women and girls of farm families represent a labor surplus and that farm incomes are especially benefited by their employment in industry—that, thus, a factory may subsidize a farm." Mississippi has just revived this plan for the readjustment period.

Another possibility, though with a sorry history, is distilling alcohol from farm products in community plants. Agrol, the original corn-alcohol experiment carried out at Atchison, Kansas, seemed to prove that power alcohol could not be produced competitively without subsidy. But stubborn investigators clung to the idea that the farmer should raise his own tractor fuel. Among them were H. F. Willkie and Dr. Paul Kolachov of Joseph E. Seagram & Sons, Inc.

The war has retarded their work but they have achieved sufficient success to revive hope that farm fuel *can* be made at a competitive price.

A competitive price is possible, they believe, if the raw materials are surpluses, culls, by-products, or unsalable grades of crops. "The utilization of these grades for power alcohol," they point out, "would mean (1) an indirect return, through an increase in price of the normal market grade by the removal of the excess material from the market; (2) a useful return, in the form of fuel from an otherwise unsalable material; and (3) an additional income from low-grade portions of the crop, the principal expense of which has been paid already by the sale of the main portion of the crop." Moreover, under their plan the farmer gets back alcohol residue for livestock feed and for fertilizer.

The cost of producing a gallon of power alcohol, not including the cost of the raw product, Willkie and Kolachov have found to be $0.0683 from a plant without feed-drying equipment, and $0.1053 for a plant with such equipment. (In the latter case, of course, there would be an added return in the sale of the distillers' dried grains.) Corn at 63 cents per bushel would bring the price of a gallon of alcohol up to $0.3051 per gallon. But the dried residue, if sold at the same price per bushel, would bring the cost down to $0.285 per gallon. While such costs will not allow for competition with conventional fuel, the costs of raw materials of no other use would allow for it. In Idaho in 1939, there were 189 million pounds of unmarketable potatoes, equivalent to 2,205,000 gallons of ethyl alcohol. If in similar future circumstances a very small charge were made for the useless potatoes, and if transportation cost to a community distillery were low, it is quite conceivable that power alcohol could be sold for less than other fuel. If the community distilleries should work like the ancient gristmills with the miller taking only his tithe, the potato farmer might well get back his next season's fuel on a barter basis. It is too early yet to discard the power-alcohol idea entirely.[3]

A suggested project similar to TVA is now exciting great interest in the nine states of the Missouri River Valley. REA Deputy Administrator William J. Neal states, "In the TVA area where low-cost electric power is plentiful, the ratio of industry load to farm load is greater than in the country as a whole. Farmers don't desert their homes because electric costs are high, but businessmen will not select high power cost localities for the establishment of plant sites when other factors are equal [i.e., accessible raw materials, transportation facilities, markets, et al.]. If America is to have broadly decentralized rural industries, it must have plenty of low-cost electric power reaching into all sections of the nation."

Even with cheap power, rural industry is not without problems. If wages paid rural labor are too low, rural industries will undersell city industries, thus reducing the ability of city people to buy the food the farmer raises. If rural wages are too high, labor might be drained from the farms. It is obvious that rural industry should not be established solely from a desire for cheap labor, but rather from a plan for using seasonal labor surpluses.

Possibilities in Rural Industry

Extracting oil and protein from cotton, soybean, castor, sunflower, flax, peanut, tung, perilla. Obtaining cellulose from stalks of castor bean, soybean, all kinds of straw, cotton, corn. Dehydration of alfalfa, kudzu, lespedeza, crimson clover, potatoes, fruit residues, green grain, hybrid corn crops, cotton, milk, vegetable residues. Fabrication of cotton, flax, wood, mohair, floss. Spinning and weaving fibres into sacks, tarpaulins, coarse goods. Refrigeration; distillation.

[3] The Foreign Economic Administration, National Farm Machinery Co., and Seagrams, working together, expect to announce soon a new engine designed especially to use alcohol as a fuel.

Vitamin and mineral feed mixing. Yard fabrication of farm buildings, equipment. And many others.

PART II

As a Means of Production

Some new farm tools are already being built on a limited scale, such as the sugar-beet harvester, which not only does the work of a half-dozen men but is threatening to revolutionize the whole industry in conjunction with other tools used at planting time and during cultivation. We have segmented seed and a new drill that plants such seed. We have cultivating and thinning tools such as the Dixie chopper, rotary choppers and thinners, and a half-dozen blocking tools, all more or less successful now and probably with new refinements coming up. These tools would have been of little value if the farmer had still needed a large group of hands at harvest, for most sugar-beet culture is dependent on huge numbers of imported seasonal workers. If the farmer did not need these workers during planting and cultivating, still there was not much use of going to the new methods because he still had to have the workers for harvest. But the success of the mechanical harvester did away with that problem, and for the future it would seem that the crop will be almost entirely mechanized.

The cotton-picker is here to stay and many of the bugs have finally been worked out of it. The few troubles that remain undoubtedly will be solved quickly when this machine can go into large production and consequent wider usage and study under varying conditions. Along with it comes another mechanical marvel, the hot-flame weed killer. If some such way as this could be found to chop the cotton, the last of the terrific human toil would be taken from the crop. The hot-flame killer works above the ground at a fast rate of speed. It sears the weeds at better than a thousand degrees and kills them with one quick passage. It's based on the principle that what little heat gets to the cotton stalk is harmless as long as that stalk has grown its sheath. The bug in the device at this point is that weeds grow like sin during the time the tender cotton plant is getting big enough to grow its protecting sheath. Thus the flame killer is not a complete success yet; but as long as so much progress has been made, it is reasonable to suppose that complete success will not be far off.

Any period of manpower shortage always brings new devices displacing permanently all manpower from some part of agriculture. This situation causes much concern to sociologists and those students who worry about the never-ending technological displacement which occurs constantly in both industry and agriculture. Here is not the place to go into that topic, but it should be noted that historically once the machine moves in, it is never permanently held back by considerations of manpower displacement. The war manpower shortage brought about a great many devices which undoubtedly will become permanent fixtures in agriculture. More, it brought many new ideas to mind, if not to complete fruition. It is likely that they will be perfected in the years immediately ahead, even though an excess of manpower does again make its appearance.

During the depression years, certainly there was a plethora of manpower on the soil. Even so, the mechanical revolution in agriculture had got started, and there was no stopping it. Thus our greatest period of tractorization, rubberization, and technological displacement went on at the very time the hundreds of thousands of unhappy Joads were thronging the highways. For that reason, and even granting a new manpower surplus, many believe further technological advancement is due in the immediate future. Here are some of the devices and ideas on drawing boards or in the model stage which show the way that advance will probably take.

The sugar-beet harvester has shown how to compensate for the varying heights to which plants grow in the row. Now it is reasonable to suppose that practically all row vegetables will be harvested by machine. The electric eye is reported in use in the model stage in California, sorting tomatoes by color so that the wanted degree of ripeness can be distinguished by the harvesting machine.

Hay, usually spoken of as the forgotten crop, probably has more men working on its mechanization today than any other item in agriculture. The new Joint Committee for Grassland Farming is expected to present many startling ideas which will take out much of the drudgery from haying while insuring the security of the product. The one-man automatic hay baler is already here and only waits on reconversion to go into large production. The forage chopper is adjudged a success and can make either chopped hay or ensilage right in the field—another blow at the old multi-man haying crew. The buck rake, in many instances homemade, is already appearing on farms by the hundreds of thousands. A worker in one of our great agricultural institutions is completing plans for a turn-over machine which will speed field drying while still retaining the leaf on the hay. Barn dryers, permitting the taking of hay from the field in almost a wet stage, are in experimental models in a number of places, notably at the University of Tennessee where the project has long been under development. It is a good wager that the weather or drying hazards will be taken out of our most chancey crop in the not far distant future and at the same time that the per ton manpower needed will drop to a low not thought possible even a few years ago.

Portable grain and hay elevators are already common in many parts of the country. These machines are cheap and are being adapted to new uses almost daily. Equipped with pickup devices, they promise to take the last of the stoop labor out of a whole long list of crops. Sorting devices, with or without the almost miraculous use of electronics, will become part of the machines. With such an improvement, grading will become more accurate than it was under the hands of long-trained and highly skilled workmen.

Mechanical transplanters were fairly successful before the war, but it was still necessary to have several men serve them. It is probable that these will be improved to the point where the plant, the fertilizer and the water will all be handled by just one man driving the machine which may or may not be self-propelled. The latter idea is causing some heavy thinking in many quarters.

At first, self-propelled machines seemed economically practical on only such big crops as wheat, for instance. The Massey-Harris Harvest Brigade gave dramatic impetus to the self-propelled idea when it moved north from Texas in 1944 doing one of the greatest jobs of harvesting in all history. But farmers often feel they can hardly justify a different engine for each machine, no matter what the virtues of the self-propelled implement. Now, however, comes the idea to make all farm machines self-propelled instead of drag or hitched in the present manner, the feasibility of the notion coming from the suggestion of making the power plant quickly and easily transferable. Thus, a farmer might make hay in the morning, plow at noon, and harvest in the latter part of the day, using many self-propelled implements but only one power unit. Some even think the tractor may be a transition machine to be displaced by this newer theory.

The workers in electricity have another idea. They see power garden tools for suburban farms, for seed growers, for hothouse people, and for persons wherever highly intensive farming on a limited area is carried on by using electricity. One idea is to have a mobile transformer set up at the end of the field with long cables extending out to operate one or more vacuum-cleaner-like power tools. Another idea contemplates storage-battery-driven tools with the batteries recharged at night when rural rates are low. Still another even more remote idea is to broadcast power to field tools. These ideas aren't all merely paper plans. Some have already been under experiment in various parts of the country and are under consideration by REA's research people.

When we consider that America does not produce enough to feed this nation as well as it ought to be fed, then ideas which look to a greater production of foodstuffs become important. It may be assumed that the doctrine of the economy of scarcity is gone forever and that people in the cities on a subnormal diet will never again be content in that status as long as surpluses are on the land. And if proper distribution gives everyone the food he should have, there will be no surpluses. If to that we add chemurgy and the need to displace depletion sources of raw materials with agriculturally produced materials, it is germane to inquire into ways and means of increasing our supplies.

The first one which comes into any farmer's mind is the tremendous wastes which are so much a part of farming. Any of us know that of all the stuff grown in the fields not more than a sizable fraction gets to the ultimate consumer. Some observers hold that only half of the stuff grown reaches the consumer. They point to cabbage, lettuce, fruit, and similar foodstuffs which begin to be wasted almost from the time they start growing. Of course, a large part of that waste comes because of the necessities of grading. The housewife doesn't want puny articles, even though they may have just as much or in some cases more vitamins, calories, proteins, and what-not than the big luscious-looking (but sometimes tasteless) top-graded article seen in the store.

Obviously, the ways to stop this wastage are two. One, do more processing, preserving, and readying-for-market in the field. Two, change the form of the foodstuff into something more esthetic looking. For the first, it may not be too visionary to see the deep-freeze process taken right to the fields. If a head of cabbage was lifted, washed, dried and "set" in its fresh condition just a few minutes from the time it was cut from the root, the saving in edible leaves would be vastly larger than it is now. Under the present process everybody from the picker through packer, shipper, wholesaler, retailer and finally housewife strips a part of the leaves—the outer leaves which have the highest mineral and vitamin content, by the way. The same holds true of a large number of other vegetables.

In fruits, only a fraction get to market because of lack of size, small skin blemishes, overripeness, greenness, bruises acquired in handling, and several other factors which do not make the item valueless. From a food standpoint, these discarded fruits may be just as good as the highly polished, carefully selected "extra-fancy" items that reach the stores. In the Western Regional Research Laboratory at Albany, California, at least part of the answer has been found. There the researchers have been able to take waste fruits, make them into a puree, quick-freeze them, and get a resulting product that looks like and has the tactual consistency of ice cream, although no cream enters the manufacturing process. At the moment, this product is considered merely a confection and during the war served as a way to get needed vitamins to overseas soldiers in an appetizing form. But further work may give fruits in their essence to millions who rarely ever get any fruit from one year's end to the next. The farmer will get a new source of income. The national production total will rise.

Thinking along similar lines, or offshoots from them, more than likely will produce a dozen other ways whereby we can preserve our present wastes for human or industrial use. Mechanical grading for size is of course fairly well advanced. Grading for color is quite feasible with the electric eye. But some nutritionists say we need even a newer grading medium, that is, by food values. They hold that the rosiest, shiniest, biggest foodstuff is not always the best or even half-best. They say that on some soils, irrespective of the looks and weight or size of the vegetable or fruit, the nutritional value is so low that the eater would be better off with a runt item from some place else.

To get that sort of grading, we must have new devices which will instantaneously make a qualitative analysis of foodstuffs in the fields. Such an idea is by no means a wild one. No doubt our skilled inventors can whip it out on short notice if the basic idea is accepted. What that type of technological progress would mean to our regional agriculture is rather frightening to contemplate. Suppose it was found that the excellent fruit of northern New York state was in fact very inferior nutritionally to runty fruit grown in, let us say, a desert out West. Suppose the public accepted this idea, and demanded a guarantee of grading by some yet-to-be-found device indicated above. Whole areas now thought to be prime agricultural regions might find their favorite crop banned by a nation of nutritionally-minded housewives.

Perhaps we had better skip over such an idea quickly, but it is well to watch the work of those little-known men, the soil chemists. They speak of some really terrible things when they converse among themselves. Many of them gaze darkly at the humid East and say it is fit only for forests and grasses, while all of our foodstuffs should come from the arid West where lime, phosphorus, and the trace elements are noticeably richer.

That the nose-in-the-furrow type of backbreaking labor is on its way out of farming anyone can see who will take the trouble to observe. Some people who still believe the myth that the tractor drove people from the soil, rather than the more correct judgment that erosion played the biggest role, look to a mechanized future and predict the direst of national tragedies. They fail to see that while the machine displaces hand and horse labor, if we are to have the better diet and better distribution we all would like to see, actually many more people will be needed in the secondary agricultural pursuits. That is, in processing, fabricating, storing and distributing agriculture's products.

For instance, as noted in the *Fortune* story, we shall need more hands in the community soybean mills. The economics of the community mill is simple. The bulk, by weight, of soybean production goes back to the farm. That's the cake, of course. The smaller product (by weight) is the oil. Obviously, it is good economy to transport the heavier and bulkier part of the production the shortest distance, i.e., from mill back to farm. The dearer yet lighter product, the oil, can then be shipped to the big fabricating centers to be made into paints, lacquers, plastics, and food products. Some of the manpower displaced from the farm by the machine can thus be utilized right in the community, contributing to decentralization, a more balanced local economy, and a healthier national picture. At the same time, that old bogey, inequitable freight rates, is licked through by-passing.

Again, more than likely cotton will be processed closer to home, with the cake going back to the farm. In the past there wasn't much sense to this, for the cotton-raising districts had little use for cake. But now that those regions are becoming our fastest growing livestock centers, the picture changes.

Quick-freezing, canning, dehydrating, and other processing phases may and probably will be decentralized and brought closer to the farm. In each case, they will make jobs for farm people displaced by machines. With cheap power and inexpensive tools a great deal more processing will be done on the farm itself. Where all of the family used to work in the fields, in a mechanized age only a few of them will be so employed while the rest will be at work in the farm's processing plant. It is cottage industry brought back by the machine, rather than forever outmoded by mechanization.

The next question is who will take care of all the present workers in centralized industry displaced by the industrialization of the individual farm, new-style? (The latter modifier is injected because we are not using "industrialization of the farm" to mean that small farms will be replaced by huge corporate-owned factories in the fields.) For one thing, many of them will be busy making the tens of thousands of individual new machine tools this change will necessitate. Another group will be expert servicers of the machines. Still another group may be absorbed right back from where they came—the farm community itself. At least one farm-equipment maker believes that instead of six million farms we shall have eight million in the not far future. He thinks he will have to employ more people than he now does to make the tools these extra farmers will demand.

Farm buildings will, of course, have to be modified to meet such changes. Already the lofty and expensive barn with its huge mow capacity is disappearing in some sections as chopped or baled hay replaces the old bulk product. There is a tendency to get whatever hay storage is used removed from the livestock, although this seems inefficient on the face of it. The fire hazard, however, is a larger cost than the higher transport. With mechanical elevators, transporters, and possibly even to come some modification of the endless-chain, moving-belt sort of thing used in industry, hay can be safely stored far enough away to give fire protection, yet transported mechanically in the quantities wanted with even less manpower than is used to pitch it down vertically from mow to manger. Such an idea is already being tried out on some farms where experiments are being made with the old litter-and-manure trolley to carry the hay from one building to another.

Another item under study at this time goes right back to the farmer's greatest of all hazards—the weather. If the weather could be modified, farming would have one of its greatest threats and greatest causes of loss stopped in its tracks. How to air-condition fields and farmsteads might be the statement of this problem.

On the face of it, the notion seems ridiculous. How could any farmer afford to air-condition his fields? Not by a machine, that's certain, but by using the very things he has long known—that is, trees, shrubs, and by diverting streams, making, or transferring bodies of water. It isn't fantastic at all when you think what we learned about bulldozers in the war: the economy with which whole mountains of earth can be moved.

Trees planted as shelterbelts not only deflect winds, they change the temperature of the air above them. Likewise bodies of water have a strong climatic effect. These, we now know, need not be kept where nature put them. If by moving them around man can advantage himself, so be it. Perhaps we shall see communities similar to the conservancy districts which will be organizations of neighbors who will get together co-operatively to move a mountain, make a lake, or plant a shelterbelt.

What good would it do? Here's an example from the writer's own community. All of us on one side of a small river have lovely lush fields of thick topsoil extending from the river bank back to a low hill which forms one side of our valley. On that black land, rich crops of truck can be grown, high yields of cereals, or thick grass with high nutritional quotient. On our hill, however, not much of anything will grow except a low grade of grass. Yet we are in an area where fruit grows well and makes a good return because of our contiguity to the city. We have established packing plants and cold-storage buildings. If we could only grow fruit on our hills, it would fit in with the types of farming we practice, make productive some pretty sorry ground, and add to our income. But we can't, although the opposite side of the valley has rich orchards.

The reason we can't isn't our poor hill soil. It is conceivable that with a smart soil program we could turn it into something more than sterile, if not the best, fruit land. But our great trouble is that in winter the winds sweeping down over those hills kill anything we set out. A very hardy type of cedar grows instead, and that cedar, to add to our troubles, is host plant to a rust which is death to orchards. Now, if we could modify that wind, warm it, or chase it higher, we might be able to grow fruit. Cedar trees aren't the answer, but were we all to get together it is possible we could find some sort of nature's windstop that would turn our side of the valley into as good fruit land as there is on the opposite side.

Another example might be an area where pyrethrum, the active element in fly sprays and insecticides, was grown successfully for some years. Then came a series of bad winters. The plant winter-killed, and the farmers, disgusted, did not go back to its culture although it promised very high per acre income. Had there been some way by concerted action of controlling the weather in that area to prevent the threat of winter-kill we might not have had a shortage of that valuable insecticide during the war.

The list of the new and exciting in agriculture is much too long for inclusion here. Only a few samples have been given. However, it can be said that scientists, researchers, and workers in many lines are hard at it all over America today. And they aren't all in experimental stations nor the land grant colleges. Many a man in a laboratory far removed from the soil is "sweating it out" to produce things which will be a boon in our Tomorrow. Never again do any of us want to see America on food rationing.

It seems to this writer that the important element in all this is not the Popular Mechanics, Jules Verne type of thinking which could be so easily adduced. Rather, it is that scientific progress, mechanization, and technological advance do not necessarily spell the end of the soil man, nor the undoubted benefits, esthetic and cultural, of agricultural life. Admitting that certain dislocation and difficulties (none of which could be any worse than those of the mid-thirties) will come to some of us, still all this advance spells the salvation of those very values we want to maintain, rather than their disappearance. Not the least of these is that the sort of agriculture we paint here will provide the antidote to the greatest of all threats to farm life—the desire of youth to go to the city. Instead, a country life which has all of the comforts of city life, plus a greater security and the traditional esthetic values, will appeal so greatly to youth that our problem may become how to maintain our cities rather than how to maintain the farm, where the greatest of all crops is grown —the Next Generation.

12. *The Atomic Threat*

WARREN S. THOMPSON

HOW vulnerable is the United States to attack from the air? I propose to try to answer this question as one who has been interested for many years in the distribution of population and, of necessity, in the distribution of industry and trade. What I have to say will be confined primarily to pointing out what kind of human and industrial targets we will present to an enemy, if there is another war within two or three decades and if the distribution of our population continues to conform to the pattern it has followed for a considerable time.

I shall not argue about the precise character and power of the weapons that may be used in the next war, if there is one, but rather shall make certain assumptions about these which seem to me justifiable in the light of the knowledge now available to every layman.

In the first place, I do not assume a third World War is inevitable. Rather, I believe that it would be folly to ignore such a possibility in the present state of world organization—or should I say *disorganization?* And if this be admitted, it would be a blunder of the first magnitude to take no measures to survive the first impact of an unannounced assault by A-bombs. I shall not discuss the political measures needed to avoid another World War, although the conclusion seems inescapable that our only real hope of safety in the long run lies in a new political organization of the world.

The chief assumptions I shall make are: (a) that if we ever are attacked again it will be an unannounced air attack; (b) that in this attack vastly more destructive A-bombs will be used than those which were dropped on Japan; (c) that the carriers of these A-bombs will be faster, will travel farther, will carry larger loads than the airplanes of this war, and, even if mechanically controlled, will strike with greater accuracy. Since the increase of the radius of destruction from about one and one-half miles for the present A-bomb to three or four miles seems well within reason, and since we are being constantly told of the increase in the radius of operation of airplanes, the improvement in jet propulsion, and the growing probability of accurate electronic control of future missles, these assumptions appear the minima that can reasonably be made at this time.

The thesis I shall support here is that the United States is, and will remain, highly vulnerable to air attack in the future, in spite of our area of almost three million square miles, if the assumptions made above are reasonable and if there is no fundamental change in the distribution of our population and our economic activity. It is very easy, perhaps quite natural, to assume that, because we have a vast territory with many of our natural resources well distributed, we are not readily vulnerable to air attack, that we are in far less danger than smaller countries with denser populations. The tendency is to admit that mere distance from possible enemies will not protect us in the future as it has in the past, but to say that our industry and people are so scattered over a vast territory that they do not present the concentrated targets European industry and cities present to aerial bombing. Hence the assumption is that we need not be greatly concerned. It is the belief of the writer that the facts regarding the distribution of our people and our industry do not justify such complacency, and it is the purpose of this chapter to set forth the facts upon which this belief is based.

If the analysis presented repeats certain statistical data contained in Chapter 3, I hope that the gain in clarity for the purpose at hand may offset the reappearance of the figures previously used in another connection.

The fact is that, while our population and our industry are spread from Maine to California and from Minnesota and Michigan to the Gulf, almost 48 per cent of our people (about 63,000,000) live in 140 communities called metropolitan districts. These districts are highly concentrated in the Northeast and along the Great Lakes, and embrace only about 1.5 per cent of our area (44,626 square miles). Moreover, of the land they occupy, only one-eighth is found in the central cities (5,478 square miles), in which live almost one-third of our people (about 43,000,000). The total area of these cities is a little less than that of the two small states, Rhode Island and Connecticut. Slightly over twenty millions of the people in these metropolitan districts live in the "outside" areas, where the density per square mile (515) is only about one-fifteenth that of the central cities (7,813), although a large number of these twenty millions live within a relatively short distance of the city limits, where the density is much greater than 515.

But to speak of 140 metropolitan districts with 48 per cent of our population understates the problem of concentration. The eleven metropolitan districts having over one million population each contain 25.6 per cent of our entire population, and the eleven with 500,000 to 1,000,000 contain another 5.9 per cent. Thus over 31 per cent of our entire population is crowded into twenty-two districts with an area of only 18,344 square miles, while the cities of 25,000 and over in these districts contain only 3,535 square miles.

This concentration of population in relatively few large communities has been going on steadily for a century and a half, and until the decade 1930–40 has shown no tendency to slacken. Even in this decade these 140 communities grew by 8.1 per cent, the remainder of the country grew by only 6.5 per cent, while the 96 metropolitan districts of 1930 grew by 28.3 per cent between 1920 and 1930.

However, the slackening of growth in the metropolitan

communities which took place in the depression years does not appear to have continued during the war. The data available on this point relate only to counties and come from estimates based on ration books. They indicate a gain of about 2 per cent in the civilian population of metropolitan counties between April 1, 1940, and November 1, 1943, while the total civilian population was declining about 3 per cent. Thus the war tended to centralize population and industry rather than to disperse them.

But the mere concentration of population in metropolitan districts is not sufficient in itself to show how vulnerable to air attack we really are. The purpose of any sudden air attack will be not merely to kill great numbers of people and to frighten the survivors into submission, but also to destroy such a large proportion of the facilities for producing and transporting war materials that resistance will become impossible. How will it profit a conqueror to kill off so many people that he will lack workers to use the resources of the conquered land effectively? The *Herrenvolk* will want sufficient slaves to minister abundantly to its most extravagant desires, but it will not be equally interested in a great and efficient industry which can assure a high level of living to the entire population. It will want only enough industry to supply it with the weapons necessary to insure the subjection of its slaves and the production of its own luxuries. For these reasons the destruction of industry rather than the killing of more than one-fourth or one-third of the people will quite probably be the prime object of the initial attack.

To accomplish the destruction of a critical part of our essential war industry as matters now stand, it will be sufficient to drop A-bombs on only a few great centers of production where much of this essential industry is concentrated. That at the same time many millions of people will be killed will not appear regrettable to the conquerors, because for the time being, at least, they will prefer a subject people dispersed into relatively small communities where there is but a small amount of essential war industry.

In order to show how highly vulnerable we are to A-bomb attack, it will also be necessary to examine briefly the concentration of our industry. In 1939 thirty-three industrial districts produced 59.1 per cent of the value of all our manufactures. These thirty-three industrial districts included only ninety-seven counties and 1.7 per cent of our total area, but they contained 35.3 per cent of our total population and almost 55 per cent of all the wage earners engaged in manufacturing. Moreover, within these thirty-three industrial districts there were ninety-four cities of over fifty thousand, with an area of only 3,144 square miles, or 0.1 per cent of our total area, and these cities contained 25 per cent of our population and over 39 per cent of our wage earners. In these thirty-three industrial districts there were, in addition, seventy-seven cities of twenty-five thousand to fifty thousand which contained another 2.1 per cent of our population and 4 per cent of our wage earners. These seventy-seven cities had a total area of only 672 square miles. They would almost certainly be completely devastated in the process of destroying the larger cities and their industry. But even these figures do not adequately measure the concentration of our essential war industry. Certain types of essential production are concentrated in a very few of these thirty-three industrial districts. In order to show this, it will be sufficient to note the concentration of the steel industry, the motor vehicle industry, and our electrical generating capacity.

The steel industry (steel works and rolling mills) had a total of 413,000 employed persons in 1939. Of this total, 37.8 per cent worked in the Pittsburgh and Youngstown areas (including Canton and Wheeling). If the workers in the Cleveland mills are added to those in the Pittsburgh-Youngstown mills, the percentage rises to 42.4. The Chicago area, which includes Gary, had 14.9 per cent more. Thus over 57 per cent of the workers in this very essential industry are concentrated in a few large mills in two relatively small areas. The fourteen largest works have almost 43 per cent of the nation's steel-producing capacity, and the twelve next largest bring the total for these twenty-six works to almost 60 per cent. The remaining 40 per cent of capacity is more widely scattered in smaller plants, but many of the latter are found in the two great steel districts already noted. Thus, much of our steel industry is not only concentrated in a few cities and their environs, but also in a relatively few huge plants.

It is true that it took repeated bombings by "block-busters" to keep the Krupp works useless, or nearly so, but the day of the "block-busters" as "blitz" weapons is past. It is no wild fantasy to say that some two score bombs like the one dropped at Hiroshima, even if only "near hits," would destroy three-fourths, perhaps more, of our steel capacity, while a like number of the super-A-bombs, which may be in use in 1960, would not only cripple our steel industry even more, but would at the same time certainly kill from 20 to 30 per cent of our people.

Clearly we need to ask ourselves whether it is possible to scatter our steel production over a wider area and whether it is necessary to concentrate more than two thousand to three thousand steel workers in a single works lying within the probable radius of destruction of an A-bomb. The question of vital importance regarding the size of steel plants is no longer, "How large a plant can be built and retain efficient production?" but, "How small a plant can be efficient?" or, "Can large plants be so dispersed that they cannot be destroyed by one or two A-bombs?" Actually our steel production is far more vulnerable today than that of Germany was up to V-E Day.

The motor vehicle industry, which produced not only the essential road transportation for our armed forces and civilians but much of the heavy ordnance actually used in fighting, is even more concentrated, both as regards size of establishment and geographic location, than the steel industry. Twenty-nine establishments with over 2,500 workers employed almost 50 per cent of the wage earners of the industry, and another seventy-eight with 1,000 to 2,500 workers employed an additional 29 per cent. Furthermore, a number of these large establishments are cheek by jowl with one another. As regards the geographic location of this industry, more than 62 per cent of all its workers in 1939 were found in the Detroit area or in near-by Michigan cities. Our motor vehicle plants, like our steel mills, are made to order for A-bombs, although the population that would be killed in destroying two-thirds of our motor industry would be much smaller than would be the case if a like proportion of the steel industry were destroyed.

Some dispersal of the motor industry from the Detroit area appears to be taking place, but the new plants being built, even though they are smaller than their parent plants, employ several thousand workers. It is significant for our purposes that they are being located near cities, most of which already have several hundred thousand people. It is certainly better from the standpoint of military security to have our motor vehicles produced in a number of plants located in different cities, even though many of these plants have several thousand workers, than to concentrate tens of thousands of workers in a small area and in huge plants near one another. It would seem the part of wisdom, however, to ask, "Can this new plant, intended to employ three thousand to five thousand workers, be broken up into several units so that it will take half a dozen or dozen bombs, rather than one, to stop all the production of this number of workers?"

Finally, in assessing our vulnerability to air attack, it should be noted that the production of electricity, while often located at some distance from centers of population, especially in the case of hydro-electric generation, is actually concentrated in a relatively few large plants. Thus the destruction of the 105 largest installations (1943) would reduce the generating capacity of the industry by 51.0 per cent. Furthermore, of these 105 plants, twenty-two are found in or near our four largest cities. The destruction of another fifty-eight smaller plants would bring the capacity lost to over 62 per cent of the total. That the loss of this power would almost certainly paralyze a large part of our essential industry does not need to be argued; nor does it need to be argued that these great generating plants would be among the first targets selected by an enemy.

The facts given above, sketchy though they are, show conclusively that our urban population and much of our essential war industry are highly concentrated and therefore extremely vulnerable to attack by airborne A-bombs. In truth, the present type of city, built in as near a circle as topographic and geographic conditions will permit, and the large modern industrial plant, built as a compact unit covering a comparatively small area, furnish the maximum possible exposure to damage by A-bombs. This is a fact which no one who thinks about the matter for a few moments will deny. It is high time, therefore, that we ask, "How can we change the structure of our cities and our industry in order to minimize the destruction which can be wrought by aerial bombing if we have another war?"

Since I do not believe that it is practicable to build any large proportion of our factories so far underground that they can withstand atomic bombing, and since living underground will be resorted to willingly only when the choice is between death and cave dwelling, I am of the opinion that the dispersal of people and industry is the only course open to us. Unless and until a world organization for the maintenance of peace becomes really operative, we may also find it necessary to build underground factories and power plants for the manufacture and launching of A-bombs. This is not as a pessimistic view, but as a realistic one.

If the wide dispersal of population and industry is the only alternative to going underground on a vast scale while a world organization is being perfected, we must then ask ourselves what form we should give our urban areas. In this we must keep in mind the desideratum that, with a radius of destruction of from three to five miles for future A-bombs, no considerable part of our people or industry must be left open to surprise attack. Very clearly this means rebuilding our cities in an entirely new form and breaking up our industrial plants into small units, so scattered and so organized that only a small fraction of essential production could be destroyed by surprise bombardment, even on a large scale.

The form of city and factory plan which naturally suggests itself is that of a relatively narrow belt of dwelling and service buildings, not more than two or three stories in height, strung along a curving highway and railway, with small factory units scattered at a sufficient distance from one another to avoid the destruction of more than one unit by any one bomb. This relatively narrow belt of buildings should not be laid out in a straight line so that any air carrier could drop bombs on it at successive intervals in the course of a straight flight. It would be better also if it were not built on a circular belt, or several concentric belts, in the case of large communities. The form best adapted to minimize bomb damage would probably be that of an irregular elongated S. If the community is built in this form, only a small part of the full destructive power of any bomb could be made effective against it; the far greater part would be dissipated into the surrounding open spaces. The exact shape of the curves used should be determined by the best technical advice available regarding the radius of destruction likely to be achieved by atomic bombs in the foreseeable future, and by a careful calculation of shapes offering the most difficult targets to airborne missles. In these new urban communities, trucks, busses, and private automobiles should be relied upon for local transport. With a good network of roads within the larger communities, and with secondary highways parallel to the main highways, not even the super-A-bomb of the future could stop essential transportation for long, since comparatively little of the roadway could be blown up or covered with rubble by any one bomb. Moreover, it would be wise to plan for alternate routes between all important points.

This arrangement would also render unnecessary the vast freight yards which characterize every large city and which proved so vulnerable in the late war. Think of what would happen to New York City if the New York Central Railroad were cut at Albany and the Jersey shore terminals and docks were destroyed! In dispersed communities no considerable population could be isolated and starved and thus rendered useless by a few bombs.

What the plan suggested here means can be made more concrete. If the atomic experts were of the opinion that the area of destruction by an atomic bomb could be increased ten times over that of the Hiroshima bomb, its radius of destruction would be increased from about one and one-half miles to nearly five miles. The radius of the belt for the new city, if the circular form is used, should then be in excess of five miles, so that a bomb dropped anywhere near the center would not reach the built-up belt at all, and one dropped on the belt would destroy only a small fraction of its circumference. The distance between the concentric belts or parallel belts, if built along S-shaped curves, under these conditions should be in excess of ten miles. It would be better, however, to use irregular shapes for all belts, so that each bomb would have to be directed individually.

An alternative to some such curving-belt structure of the city would be to break all cities of over fifty thousand or one hundred thousand into smaller cities, containing perhaps not more than twenty-five thousand people. Such small cities should then be scattered over the countryside at distances from one another great enough to prevent the destruction of more than one by any single explosion. Besides, the factories in these small cities should be as small as is consistent with efficiency, thus to avoid the risk that any considerable fraction of an industry could be destroyed if a particular city were obliterated.

The dispersal of population and industrial plants into small units may or may not mean the loss of a significant part of our industrial efficiency. This, however, is secondary to the protection of essential producing units at a time of crisis. The fact is we are no longer free (at least we are not until we have a world government which can do a thorough policing job of all peoples at all times). We cannot choose the most efficient economic structure just because it is efficient, or because we like it, if it does not also guarantee us sufficient immunity from a surprise attack to enable us to launch an effective counterattack.

Unless we develop an efficient world state, our choice lies between organizing a new social and economic structure which will yield the largest possible measure of safety in the event of a sudden aerial attack, and the enslavement or destruction of our nation through the bombing of our great cities and industrial plants. We have no more choice in this matter today than we had as to whether we would fight Japan after Pearl Harbor. No one now doubts that the Japanese war lords would have made slaves of us if they had had A-bombs available in 1941. If our power to fight back is destroyed in an initial attack, we must either submit or be killed in futile revolt.

The failure to rebuilt our urban industrial communities to withstand aerial attack while we are striving for a world government which will have the power to control atomic energy may well make us, or at least those of us who survive, the slaves of the "master race" which plans and launches a surprise attack. Even if we have prepared an equally effective counterattack and are able to launch it, this will not save the people or the production in our cities. The instantaneous slaughter of one-fourth to one-third of all our people and the destruction of one-half to three-fourths of all our industry is the price we may very well pay if we continue to crowd people into the present type of city and build only large factories. The simple fact is that a few hundred atomic bombs like the one dropped on Hiroshima would largely destroy our great cities and our essential industries. A few score bombs having ten times the area of destruction of the Hiroshima bomb would destroy all of our cities of over 250,000. Only the wide dispersal of our people and our industry will give us a chance to survive a surprise attack and retain a social order which would be worth living for. A new element has entered human affairs with the development of the A-bomb and the long-range aerial carrier, and we can ignore its implications only at the peril of our destruction as a free nation a few years hence. If we take no measures to minimize the havoc the new weapons can wreak upon our present urban communities, they and their industry are in constant danger of obliteration. This is no fantastic imagination; it is, unfortunately, a sober fact which we must face.

I know that it will be said that such dispersal of urban population as has been suggested here, and such a fragmentation of production as would be necessitated by this dispersal of people, are too costly to contemplate. We would have said the same in 1940 about a war which would cost us 250 billions. But I know of no one who wishes that we had saved this money by allowing the Nazis and Japanese to impose their will upon us. No cost is too great, if it can be borne at all, which will save for us the essentials of a way of life we believe worth while. I can only repeat again that, until and unless we have a world state which can police all nations effectively, we can only choose to continue our present economic structure, based on the modern large urban community with great productive units, at the risk of annihilation at no distant future. Today we are free to live in great cities and produce our goods in great factories, just as the armored knight with his lance was free to continue his type of warfare after the foot soldier appeared on the scene with a firearm.

A Cervantes of 1960 will be able to make our present-day behavior in the face of the A-bomb and the airplane appear as absurd as the behavior of Don Quixote, if there is anyone left to read him. The only sensible thing to do, of course, is to perfect a world government and endow it with the power to police all the nations of the world. But this will assuredly take some years. In the meantime a policy of dispersal of population and industry vigorously applied is the only means of gaining for ourselves a small measure of safety, because this action alone lies entirely within our own power.

But dispersal, like any other defense measure, is merely a temporary solution. It cannot, in the very nature of things, provide the ultimate security for which men everywhere are searching. I do not offer it as a substitute for a larger and more enduring means to man's welfare, namely world organization. Nor do I assume that it will prevent vast destruction of life and industry in the event of any surprise attack made upon us. Dispersal offers a considerable measure of safety while the greater measure is being perfected. World organization can and must be achieved.

13. *Moral and Cultural Aspects of Decentralization*

ROY L. SMITH

THOREAU, quoting Damodara in *Walden*, remarked that "there are none happy in the world save beings who enjoy freely a vast horizon." Upon the plain, in the mountains, or by the sea, life has a broader meaning for all save those who, from early life, have been robbed of their feeling for the world of nature of which they are a part. It is impossible there to avoid spiritual implications which are all too often never met in the larger communities that man has built for himself.

However, it has been objected that, in exchange for the freedom and the spiritual values upon which man can achieve in less crowded communities than many we know, he gives up many of the advantages that urban civilization has given him. This might have been so in a more primitive age—as a matter of fact, I am not sure that our urban civilization was not a necessary step in the evolution of something better. Music, the arts, literature and drama have been produced upon the urban stage—Athens, Rome, Florence, London, Paris, New York. Scientific achievement, hospitals, public health, have been encouraged and sustained by the metropolitan centers, past and present. Bryant said, in "A Forest Hymn," that "the groves were God's first temples," and no doubt that is true, but no one would take the trouble to question the prodigious function, in religion and morality, of the great cathedrals of the cities, the teaching nuclei, the compact mass ethics of metropolitan centers which are inextricably interwoven with music and art. To mention just a few outstanding symptoms—Leonardo's "Last Supper," Handel's

"Messiah," Michael Angelo's "Moses"—is to suggest an infinitude of items. We have to remember, also, that all of the great modern universities were established as a part of religious programs, hospitals were first promoted by religious groups, and the very art of printing itself came about because of the efforts of religious teachers. In this interweaving of the cultural and moral backgrounds of the race, we plainly see the importance of the city grouping.

Yet to accept the theory that what has been must now continue indefinitely under identical auspices is to confuse an old fact with a new reality. Today the fetters which once bound us within city walls are no more. It is quite as possible to have a theater in a small place as upon Broadway. Has it been forgotten that, because the legitimate stage failed to move with the new transportation at its disposal, Hollywood was raised up from a wheat field?

Hospitals and medical research need not be confined to the metropolis, and probably will not be. Science itself has brought the arts conveyed by radio and television to the rudest hamlet in the land. As Sainte-Beuve once observed, we too often merely adore the fact without being able to interpret it. We speak of the annihilation of space and time, but we seem unaware of what this really means to us in terms of advantages, cultural, intellectual, moral, and physical, particularly as they affect our mode of life.

It is hardly more than a coincidence that the wealth required to sustain cultural and scientific advances in the past was concentrated in the cities. Who can say that the process may not be reversed in the near future?

It has often been said that civilization contains within itself the seeds of its own destruction. This theory is implicit in the writings of Mumford, Spengler, and other exponents of the cyclical theory—summer, autumn, winter, then spring again. While it is not necessary to accept the gloomier aspects of such thought—unless we admit the possibility of the destructive use of atomic energy on a wide scale—it remains that the law of diminishing returns is already at work upon the city. This has been made clear for the economic and social levels in earlier chapters. The aspect that concerns us here is the life of the mind and spirit. The fact that the big city has nurtured intellectual and artistic impulses with vast usefulness does not argue that it will increasingly perform this function as it becomes larger and larger and more and more congested.

A discerning study of the life of the American people indicates that the trend of diminution had begun some time ago, and that the rise of our great cities has been accompanied by a definite decline in our spiritual life. Inasmuch as the basic strength of a nation is derived from its spiritual, rather than its economic resources, this means that further growth of the great centers of population will further undermine the nation's vitality.

This study is not an appeal merely for the support of organized religion as it is represented in the United States by the church and the synagogue. Rather it is an effort to set forth certain fundamental considerations which must be given the most careful thought by those Americans who love their country and aspire to its best fulfillment. The preservation of the spiritual life of the nation is the highest form of patriotism, is it not?

That this proposition is not based upon vague, stratospheric notions is substantiated by such a hard-headed, tough-minded man as Chief of Police L. J. Hilbert of Oklahoma City, who has publicly stated that only 3 per cent of the juvenile delinquents appearing in police court have had religious instruction. No one can dismiss this fact, much less at a time when juvenile (with implied parental) delinquency is becoming a national problem.

It is one of the elemental facts of human experience that the human spirit demands something which it can worship.

This means, in simple language, that every man has to believe in something bigger and finer than he is in order to endure life. Millions find this "something" somewhere inside, or in organized religion with its vast variety of institutions and interests. Others find it elsewhere—in lodges, philanthropy, reform movements, patriotic activities, social enterprise—and live with abandon in behalf of their "cause." They are loyal. But so is the underworld, which has its code, for the violation of which many a hoodlum has paid with his life.

Powerful forces are operating in every great center of population to disintegrate our loyalties. The process begins when a baby is born in a huge metropolitan hospital and at the end of a ten-day period is brought home to a rented apartment. The chances are that before the child has finished the elementary grades in the public schools he has been moved three, four, five, or even more times. I know of children whose living place has been shifted no fewer than six times before they were graduated from high school. It is inevitable that such children will experience a serious loss of the sense of stability in life.

Psychologists agree that the growing child is extremely susceptible to the influence of his surroundings. During those first almost unconscious months he feels secure in his mother's arms; they are, to him, the symbol of the security and stability of the universe—in fact they are the universe for him. Later he stretches his reliance to include the room or the house in which he is growing up. Within a few years he is living in a neighborhood, but his home remains the center of his world, and to it he flees in time of emergency. Careful studies made among thousands of British children who lived through bombings reveal the significant fact that children do not resent the destruction of property nearly as much as they do the loss of familiar landmarks.

Exhaustive psychological studies in wartime reveal the shock a child experiences in being transplanted from one dwelling to another, even when the physical conditions provide adequate comforts. The child who is moved from one apartment to another is under the periodical necessity of making some of the most fundamental adjustments of which the human spirit is capable. This means that serious damage is done to the tender roots of life which are constantly being put down and torn up. It is a rare tribute to the divine quality of their spiritual lives that children are able to survive such transplantings. Certainly nothing else in nature would. Contrast the spiritual experiences of a child who has been born and reared in the country, and who has lived under the same roof through the first sixteen years of his life, with those of a city child who has been moved from one apartment to another, perhaps from one school to another, over and over again during the same period. Real-estate figures in great cities bear eloquent testimony to the restlessness of urban dwellers. The erection of one twenty-four-family apartment building will result in the movement of at least one hundred families. In fact it almost might be said that the existence of a big apartment house is in itself prima facie evidence of a shifting family life.

As the sense of stability diminishes in the life of a child, his sense of loyalty also diminishes. Whereas the rural youth will point with some inner pride to "the place where I was born and brought up," the city youngster might have to point to half a dozen street addresses. In no apartment did he live long enough to develop a personal affection for his surroundings. If he has changed schools, the shock has gone deeper. These delayed or destroyed loyalties, which are part of the vengeance the city is heaping upon the nation, are all too serious to be ignored. A child who grows up without developing loyalties to the house in which he lives, the school he attends, or the crowd he runs with, will become an adult citizen incapable of a great possessive loyalty to the nation which protects his life and property and provides for his opportunity.

The child growing up in the country learns to assume responsibility for the care of property. If a window is broken, he knows who pays for it. If the house is heated, he knows who brings in the fuel. If repairs are needed, he knows who makes them, and may assist in carrying them out. All this has the effect of making him cautious in his treatment of property, and as this sense of responsibility develops, other responsibilities fit into their natural places. Perhaps minor tasks are assigned to him; in the process he discovers that life is a matter of co-operation among all parties. Father brings in the food, mother prepares it, and the boy brings in the wood. If there is no food there is no dinner. Everyone depends upon the faithfulness of all the others, and in this dependence we have the germs of a stable national life.

The foregoing is a homely application of certain truths that seem fairly self-evident, but may they not be the core of a vaster and more profound set of values, extending to and permeating the whole fabric of our society?

The depression years revealed the extent to which we, as a people, had become accustomed to the thought of allowing the state to take responsibility for the individual. Preachers, uplifters, politicians, and courageous editors have lifted up their voices in defense of the rights of individuals which constitute the pride of democracy. But the time has come when there is desperately needed a revival of the sense of individual responsibility, and at the same time we find the American city drying up that sense inside the souls of hundreds of thousands of individuals.

There is something overpowering about a great mass of humanity. The very crowds upon a city's streets suggest to the individual his utter incompetency. His protests are so inadequate; his opinions are given such scant attention; his sufferings provoke little sympathy; his loneliness attracts little notice. If he is a worker he feels miles removed from his employer. The likelihood of bitter class conflict increases in direct ratio to the mass of workers, because of the disappearance of the close personal touch between employer and employe. If the metropolitan dweller is a voter, he is separated by an almost impassable distance from "the administration," particularly if the governing body is a large bureaucracy in a centralized scheme. If he is a consumer, he is separated by a great gulf from the producer, and in the logistics of a supposedly peaceful working world, his supply line is lengthened to the point where he becomes almost frantic with anxiety for government-conferred security. All this has the effect of destroying in him all sense of his importance as an individual. The only way he can hope to count is by becoming a member of some organization. This means pressure groups. It means that tenants are arrayed against landlords; employers and employes do not live in "one world." The bloc is the refuge.

It is inevitable under such circumstances that the individual loses confidence in himself as an entity; in turn his self-respect is assailed. "What can one man do?" is the despairing wail of millions who see the evils all about them and despair of ever being able to do anything to improve conditions. From despair it is but a short step to indifference and cynicism, and that is sufficient explanation for the fact that the average great American city represents the most inefficient and unsatisfactory government in the world.

It goes without saying, of course, that there is a solid core of decency at the heart of every great city. Except for that fact such great centers as New York, Chicago, Los Angeles, and Detroit would have destroyed themselves. When the "curfew order" was issued in 1945 by the federal government requiring all night clubs and drinking places to close at midnight, the Gallup poll disclosed that 95 per cent of the people, on average, were in bed an hour earlier. But the headlines, the radio comments, editorials, and general conversation all centered upon 5 per cent of the population. This is typical of urban morals—the minorities get the publicity and have the effect of setting the standard for the majority. To a substantial extent, community morals spiral under the impact of the 5 per cent and their obscenities.

It could be true, as some contend, that there are just as many evil persons per thousand in the country as in the city, but the difference appears in the simple fact that antisocial persons get together in the city, and their numbers encourage them in their evil.

An additional factor encourages profligacy in urban environments. An individual who is restrained from evil living by the force of community sentiment in a town of ten thousand is relieved of that pressure in a city of two million. It has been shown by actual case histories that thousands who would be numbered with the respectables in the country slip easily into the ranks of the disreputables in the cities.

Dr. Carl Jung, the psychologist, declares that the majority of those who suffer from mental derangements have previously lost their religious faith. There seems to be a definite connection between a vigorous religious faith and mental health, according to the testimony of other authorities equally competent. As the pressure of the city destroys faith it undermines the forces of religion.

From every survey this writer is able to consult it becomes apparent that the churches are gradually losing their hold on the city dwellers. Many charges of incompetency can be made against the church, and proved, but thus far no satisfactory substitute for the church has been found.

With a welcome turning from a destructive to a constructive thesis, we may contemplate the possibilities of creating a climate in the smaller cities and towns for faith, science, and culture, which, in ages past, have sprouted miraculously from religious backgrounds and inspirations. To say this is not to set up a special pleading for religion—it is merely to say what happens to be a fact. We all know how learning and the arts were nourished through the Dark Ages in the monasteries, and how our greatest universities were founded.

In education and the arts, some of the very best accomplishments have come from the smaller places or rural areas in America. Nearly every state now has its art center, and from other than metropolitan areas have come such significant painters as Grant Wood, Thomas Hart Benton, Birger Sandzen, Oscar B. Jacobson, and John Steuart Curry. Museums like the George A. Joslyn at Omaha and the William Rockhill Nelson at Kansas City exist, and undoubtedly more of them will be created within our lifetime—in places where they will accomplish the maximum of good. As someone has remarked, to deny the great body of Americans the cultural benefits of the theatre and of the graphic arts may be compared with a situation in which the children of a metropolis are granted the very best in educational advantages while those outside are told that they will have to get along as best they may.

Letters, fortunately, are tied to no very definite material conditions. If America has a literary tradition, it is not based upon a purely metropolitan concept. As a matter of fact, literary strength has derived from a wealth of "native" themes, the importance of which has become abundantly clear within the past two decades. The increasing strength of regions in this respect is significant in itself, but it may also prove that, if universality is to be attained, the large will have to be mirrored in the small. Life simply cannot be represented in a vacuum—it has to be placed in time and locality and made understandable in terms which recommend themselves to humans. It is futile, therefore, to expect that writing will abide by the rule of the typical—if the metropolis, which has established itself as the norm for so many other things, can be said to be typical.

When we speak of "typical American music" or "typical American art," just what do we mean?

It seems that this question and others will have to be answered before we can really weigh the significance of cultural decentralization.

If we should go by a counting of noses, we should discover that most Americans live in cities and towns and very few live in the country. Then if we assume that the big town calls the tune for the little town, Tin Pan Alley wins by a huge majority. As a matter of fact this is precisely what has happened. But it is a poor aesthetic that measures cultural values by mere majority rule. If such a principle were valid, most of the really worth-while cultural institutions that we have inherited from the past would have died aborning. If the majority has often rejected the musical accomplishment of the moment, it is hardly less true that what the majority has often approved has not survived beyond the generation which produced it.

In short, it is absurd to assume that the vast majority of the musical compositions which are produced and played these days are either good or typical of enduring American taste, merely because a preponderance of urban dwellers approves them here and now. But in this statement there is no implication of either rightness or wrongness: tastes simply do not admit of such absolutes. Yet it is impossible to avoid the conclusion that, at any given moment, our education, art, literature, and music reflect in considerable measure the values we attach to life. In twentieth century America, we are beset by materialist thinking, and if dissonance in music is the rule, we ought to know why.

It is better on the whole to say that Tin Pan Alley typifies us at the moment, than to say that swing is "typical American music." The long roll of history will give a surer answer: if an urban-materialist civilization prevails, swing or something resembling it may be our abiding musical expression; if an order closer to the ideas presented in this book emerges, then we may expect something else—I think a better "something else."

On the whole, it would seem that the encouragement of regional cultures, in decentralized environments, would provide vast new outlets to creative arts that now seem to be "cribbed, cabined, and confined" by steel and concrete urbanism. Considerable emancipation may be anticipated when culture finds its way naturally among human beings who live close to the soil, and there should be a great and significant freshening of current. If a New York publisher wishes to find out what is "typical Arkansas," he should get a referendum in Arkansas, not on Lexington Avenue. And it is worth while remembering that the Lake Poets of England were regionalists, yet they spoke a universal language. There are many analogies in America.

14. No Blueprint for Utopia

ELMER T. PETERSON

THIS is a case in which the word "no" does not necessarily mean a negative approach. One cannot blueprint in advance the innumerable rootlets that establish a tree, or the clouds that bring nourishment of rain to unpredictable plants or human beings.

This book is no promotional tract for any cult or ism, hence it avoids blueprints. It is written for those who believe we can move forward within the framework of our present society without painful, wrenching dislocations or tensions, and bring about an orderly decentralization under a diverse pattern.

Some enthusiastic advocates of decentralization seem to believe that it is necessary, in order to achieve what they consider to be the ideal objective, to adopt Henry George's doctrine, or nationalization of land, or a radical change in the monetary system. Without attempting to assess such theories, the editors of this book feel that it is sufficient simply to suggest various approaches to the evolutionary problem under auspices of individual initiative, encouraged by such a governmental policy as is likely to create the appropriate climate and ecology for decentralization.

In fact we feel that Utopia, if such there ever will be, is necessarily unblueprinted, whereas a centralized society is one with the most blueprints.

Whatever we may think of Ralph Borsodi's specific recommendations, we can probably all agree with his point that we need a new outlook on education—a vigorous reorientation. We can sympathize with Josh Billings' sage remark that there are many wise people in the world, but the trouble is that "many of them know too many things that ain't so."

Gathering together some of the salient points in previous chapters, we may enumerate various items of folklore that have crept into our ways of thinking—folklore that may be seriously questioned—namely:

1. *The saga of bigness.* Up to the present time there has been a tendency to accept, almost without question, the theorem that bigness is in itself virtuous. First we worshipped the bigness of the country, the Great Lakes, the Grand Canyon, the Rockies, the giant redwoods. Well and good. Then we unconsciously transmuted this worship to man-made things—the biggest wheat crops, the biggest factories, the tallest buildings, cities with the biggest population.

If you analyze the bigger-city complex that exists in virtually every community from Post Oak, Missouri, to New York, you will find that a large and influential element of the community hopes to profit by the unearned increment that accompanies congestion. This is true not only of the real-estate operator, the capitalist, the merchant, and the manufacturer who hopes for a bigger market, but also of the lawyer, doctor, home owner, minister, teacher, day laborer—in fact persons of every conceivable occupation and status. Everyone has a vested interest in expansion and congestion. Part of this vested interest is justifiable—even admirable, in that it stands for courage, foresight, and hard work. Since practically everyone has a selfish interest in wanting a bigger home town, it is no wonder that the bigger-city complex has become a ruling passion and a national cult and folklore.

Through all the hurly-burly of the bigger-city movement we can now trace the fact that there is really no object in continuing the cult, for the national population curve is flattening out, and the bigger the population, the greater the race suicide.

Moreover, as said before, the modern facts no longer argue for bigness in many of our most important institutions. We shall have to reconsider our folklore of growth and set aside for special study those areas in which bigness is a positive detriment. Big capital, big labor, and big government have had their innings. They have shown seamy sides, with some of the seams bursting.

2. *The legend of efficiency*. The trouble with this particular item of folklore is that it has arbitrarily constructed a false set of values.

Picture the familiar big business executive in his metropolitan office. With three telephones at his elbow and several secretaries hovering near, he gives curt orders and makes things move across seven states. For two or three hours he gets things done. We stand in awe of this very efficient man. But perhaps he spends another hour or two in a conference at which yesterday's bridge game or the difficulty of getting good whiskey may be the prime topic. He makes his luncheon do double duty as the occasion for provender for the inner man and as means to working in a conference that the business day will not permit. He commutes the long way home and is aware when he gets there that exercise, which was a habit in school and college days, is out of the question. He may (and many of his kind do) develop heart trouble. If he fails to attain normal life expectancy, the doctors and actuaries will know why.

Efficiency is not a game in which a man tries to outdistance the clock and the machine. It consists in achievement without disastrous expense to the human mechanism. In an age which is rapidly developing machines to take the place of man power, it is obvious that the best use to which we can put man power may be in occupations, avocations, and diversions which do not produce direct material or pecuniary profit, but intangible reserves of the mind, health, and spirit.

3. *The cult of statism.* The growth of reliance upon central government, not only for the customary political and legal guarantees in a democracy, but for very large areas of economic and social action as well, is a phenomenon of the twentieth century. The tendency is, of course, an outgrowth of the constantly increasing complexity of industrial civilization. But from a sound interpretation of the national state's capacity for making adjustments and resolving conflicts we have come to assume naïvely that the executive and Congress can in fact do all things for all men.

Among sensible people, there will never be disagreement over the proposition that the task of government becomes larger and more complicated as society itself becomes larger and more complicated. Yet this is not the whole truth. Central government becomes a mere handyman if it is not able to claim two large resources: the self-reliance and self-respect of individuals, and the generative powers of local units of government, the state, the municipality, the town, and the county. It is here that democracy, working towards a full life for its citizens, can be made to function in a way that a centralized system never can. It was not without purpose that the Founding Fathers gave us a federal plan of government.

Big government, like big everything else, encounters sooner or later the law of diminishing returns. It is little short of stupid to think in terms of economic and cultural decentralization without reference to a corresponding decentralization of government function.

4. *The myth of the farm as service station.* The industry of agriculture is just that—an industry. Biologically, however, it comes first. If we are not quite aware of this, the people of the devastated countries of Europe and Asia are. As centralized industrialism has grown we have adopted the theorem that the farm is the service station for urban dwellers. One corollary (quite as false as the original proposition)

is that people should move off the land into the cities, because mechanization has made it possible to produce cash crops with less manpower than was formerly the case. Machines take the place of manpower in the city as well as on the land. The erosion of farms is inseparably linked with the erosion of the farming population. And every student of farm management and food production knows that the huge farming corporation is not the solution to the problems of conserving soil and providing an adequate food supply for the people.

5. *Interdependence is in itself virtuous.* This fallacy has gained unaccountable prestige, possibly by mere repetition. Of course we know that interdependence is dependence multiplied by ten or ten thousand, as the case may be, and is correspondingly opposite from independence. Interdependence may be inseparable from large, complex organization, but it may be a necessary evil in many cases, rather than a virtue, since nothing has happened or is likely to happen that will dim the luster of independence, which is a natural concomitant of decentralization.

6. *The theorem that, if large aggregations of specialists are good, larger and larger aggregations, to an infinite degree, are still better.*

A corollary is the theory that specialization is always the most efficient way.

These notions require thorough exploration. They carry both plus and minus values. Excesses are often dangerous.

There can be no question about the value, for example, of technical engineering training as offered by urban centers. Such training is 100 per cent plus value. Even so, the testimony of a motor engineer who held an officers' commission in World War II regarding the handling of motors, including trucks, jeeps, bulldozers, planes, and power boats, on many battle fronts is significant. He says that the average farm boy, accustomed to operating and handling a wide variety of implements, was always superior in a tight place, when repairs or adjustments had to be made quickly, perhaps under fire, because he had acquired a resourcefulness and an independent power of thought that was not so evident in the case of boys trained exclusively in the urban technical schools. Carried over into the difficult field of peacetime economics and industrial life, the same resourceful lad has a better chance of success than the narrow-channel specialist.

When carried to extreme, any specialization is likely to become a form of ignorance, since it fails to correlate with the broad field of human problems.

7. *The failure to differentiate between durable satisfactions and sensory or neurotic stimuli as goals to be attained.*

There is no need here to lay the ghost of hedonism, or to draw too fine a line between "satisfactions" on the one hand and "pleasures" on the other. Socially, however, it is admissible that certain pleasures are to be encouraged on cultural and spiritual grounds, while others are not.

Teachers College of Columbia University, the State College of Pennsylvania, and other institutions have made gathered subjective evaluations of happiness, and their grand average finding is that durable satisfactions, which produce genuine happiness, are found in natural environments or in simply "talking with other people," and not in drinking, dancing, gambling, going to night clubs, or taking part in other typically urban diversions, which rank lowest. These studies do not even touch the moral phases of the problem.

We cannot claim to be thoroughly civilized until we make clear differentiations and have clear ideas of what goals we desire, and why. Granting that sensory pleasures have their rightful place, a truly educated man must put each type of pleasure in its proper category, even when he employs an amoral approach and then acts accordingly. Many people are in a more or less constant daze of disillusionment, like a bird trying to peck at berries through a plate glass window, because they are constantly frustrated in their search for true happiness. Certainly we need a new point of view by education on this point.

To attain re-orientation on the foregoing notions and other related ones, one need not be an ascetic Thoreau. He may adopt the golden mean, granting full credit to the splendor of human invention, imagination, and acquisition of special skills, yet ceasing his bumptious fight against nature, resolving that he will go along with her in an informed spirit.

This book is not a special pleading for a purely rural existence. Rather, it has tried to show that industry, to be healthy, must be soil-born. As related to the employment problem, then, the proposition may be stated thus:

"There can be no distress man-power surplus on the self-sufficient, live-at-home farm; conversely, any chronic urban man-power surplus has no choice but to become parasitical."

If we accept this approach, we immediately see that a rational economics, with all its social and political implications, is built from the ground up, and not from the top down. The average doctrinaire is hopelessly infatuated with schemes that can be thought out in the seclusion of a study, blueprinted and imposed on the masses by the *Herrenvolk*. That is not the way our extraordinary American civilization has been constructed, nor will it be the way it is reconstructed if we keep our wits about us.

There may be some special significance in the fact that Detroit, which led the way in modern mass production and low price techniques, thereby giving the nation its biggest impetus towards industrial prosperity, is also furnishing leadership in decentralization, with such great organizations as Ford and General Motors giving support. Moreover, the Detroit Board of Commerce, in an important meeting March 1, 1945, adopted a new charter for prosperity and welfare under the title, "Prosperity From the Ground Up," which boldly and unreservedly declares that agriculture is the starting point for all industrial welfare.

The pronouncement puts soil conservation first on the list of things to be done. Then come improved marketing, small farm mechanization, better health conditions, and other desiderata. Throughout the declaration there is an unreserved expression that one of the greatest industrial communities of the country owes everything to agriculture, and that the chief elements of prosperity are built solidly on the soil. Henry Ford's rural factories in the Dearborn region, where men may work part time on their small farms, have been successful and inspiring. Roger M. Kyes, the president of Harry Ferguson, Inc., designers and distributors of tractors and other farm implements, gives the pitch for the new song when he says: "The big farm produces crops—the small farm produces men." For that reason the greatest emphasis will be placed on mechanization for small, live-at-home farms, so that the small farmer or the industrial worker with one foot on the soil may be relieved of the drudgery of those types of chore farming which produce the crops most needed —fruit, vegetables, milk, and poultry.

Without blueprinting the future, it is nevertheless impossible to escape certain tendencies that are bound to affect it. Transportation (which we say has annihilated space) will be one factor. What precisely does it mean to us?

A second world war was required to prove to us that rail transport, which might have lifted us out of the trough of centralized industry much earlier, was badly in need of attention. We knew all along that the system of rates arranged for the country as a whole under the sponsorship of the Interstate Commerce Commission was working against our best development. "Official territory," inequities, the failure to adjust the whole scheme upon some rational, equitable basis were allowed to continue in spite of the lessons learned in World War I. Now we know more fully what it cost us in money and in delays to continue such a system during the most critical production period of our national history. As a matter of fact, the cases now pending against railroads are an indictment of the Interstate Commerce Commission itself, whose rate-making philosophy, to quote the very mild language of a Supreme Court justice, is "empirical."

There can be no doubt that the rail-freight problem will be solved within a relatively short time. When the inequities have been eliminated, the areas possessing resources will unquestionably move into the fabrication of commodities, instead of shipping the raw materials into "official territory" for manufacture there. Hundreds of small communities can be placed upon a sound economic basis, and markets are likely to be opened which were hitherto dormant, partly because of the high costs of commodities shipped in from a distance, partly because of the inertia of our distribution system.

Air transport is upon us. No one can yet measure the transformations that will affect our economic scheme, but it seems perfectly obvious that reasonable rates and the speed of delivery of commodities of all kinds should result in the breaking of some of the barriers that have held many areas back. Cheap atomic energy for industry is next.

The telephone, telegraph, radio, and television are other instruments to the larger purposes envisaged in decentralization as proposed in this volume. They, with other forms of communication, transportation, and the still to be developed resources of science, must make for a far more fluid concept of social organization than we have been willing to admit in the past. If this can be said of the material aspects of our culture, it can hardly be denied for the spiritual side: as we really settle the continent which we have hitherto claimed but not developed, the rich cultural manifestations of our civilization should be more widely shared.

The responsibility of the manufacturer and his advertising agents has in this country always seemed appalling. Nowhere else is it necessary to conceive success in continental terms. To create a better mousetrap is one thing, but to give it currency from Maine to California is quite another. In one sense this apprehension is the result of an admitted fact, that the United States of America is vast, heterogeneous, and moved to action only at great cost. In another, it is the result of a vaguely felt knowledge that the job really can't be done after all—at least not as well as it could be done under other circumstances. And those circumstances have to do with decentralization.

In spite of everything we have not wanted to do, either for selfish or for imagined rational reasons, the movement towards decentralization has been under way in this country for more than a generation. By sheer necessity regional cohesion has begun to take form—on the Pacific Coast, in the Southwest, to a certain extent in the South (though there never has been any doubt about political cohesiveness there), and in the Middle West. Economies built partly upon the "cluster principle" embracing several states have begun to emerge. The movement has been marked, but it has been retarded by a number of factors: inertia, the conspiracy of the freight-rates system against it, the lack of or indifferent application of capital and special skills, and the failure of industry based elsewhere to embrace the tide of the times.

While economists and political scientists have always recognized the lag between the fulfillment of one condition of society and the coming into operation of another that is to succeed it, the disparity here is a good deal more serious than we may know. It may even account in considerable part for the admitted hazards that we face in the next ten years, if we make no effort to decentralize. I am far from believing that every man has an innate desire to live with one foot on the land, but I cannot overlook the salient facts that Jennings J. Rhyne has presented earlier in this book with reference to North Carolina. Fortuitously, that state has spread the risks for its inhabitants long before the great depression appeared. Its economy was so broadly based, it had drawn so happy a relationship between land use and industrialization (in relatively small centers), that its depression record (statistically at least) was better perhaps than that of any other state.

Oklahoma recently discovered an interesting fact about its position which may illustrate, by a *reductio ad absurdum*, the *un*economies we are practicing rather broadly in the land. Governor Robert S. Kerr, addressing the Southern Governors Conference at Biloxi, Mississippi, said:

"We found one county in Oklahoma where peanuts were produced on the farms and glass jars were being manufactured at the county seat. Both were being shipped to another state where the peanuts were processed, put in the glass jars which were then labeled, shipped back to Oklahoma, and sold in the county in which both peanuts and jars had been produced."

Transportation is friction in the economic machine. It represents waste when any given section is self-sufficient with respect to any given type of material and proportionate consumption of that same material. Long hauls may be necessary to get producer and consumer together on an exchange basis, as when Boston requires western leather, or Nebraska wants codfish, but a round-trip haul for any commodity is indicative of economic malaise.

Probably there is no compilation of statistics showing the total volume of wasteful freight traffic of which the peanut transaction is but a tiny symptom, but even a casual survey shows an enormous amount of aimless hauling back and forth of bulky commodities—carrying coals to Newcastle, leather goods to Texas, corn products to Iowa, and breakfast food to Kansas. If we believe with the majority of economists that production is our real wealth and that transportation adds to the sale price of a commodity (often without rational justification), we are forced to the conclusion that unnecessary hauling annually subtracts an enormous toll from our true national wealth and deflates much of our national saga of efficiency.

Some realists may cynically admit this vast waste of man power and machine energy, and then inquire: "What can we do to keep men and machines busy, instead of hauling things around the country?"

One realistic answer would be to put the men and machines to work saving our basic wealth—the soil. A ranking official of the U. S. Soil Conservation Service has made a rough estimate that one state alone would require several hundred million man hours of labor, with corresponding use of machines, to stop the erosion of the topsoil (of which one-third is already gone into the ocean), with additional hundreds of millions of man hours annually for maintenance. It should be remembered that when the soil goes, so also goes urban industry, business, the professions—everything.

In addition to the illimitable new frontiers suggested by atomic energy, synthetics, plastics, television, and an almost infinite number of other scientific and industrial items, we have to consider the new frontier of civilized ugliness which was created by the ill-advised "conquering of the wilderness." In the badly managed original "conquering" there was criminal exploitation, not so much by farmers, individual miners, trappers, or small-scale timber-cutters as by urban-based cream skimmers who had little respect or affection for the land except with regard to its cash output. Our prodigious and challenging task is to cure the old scarred frontiers by healing the broken sod, saving the soil, rebuilding the land with humus, organic micro-bodies, and chemical fertilizers, and by bringing the human race more nearly into a sane and balanced relationship with nature by reversing the stampede to the metropolis.

It may be necessary to remind the American people that a colonial system of exploitation which, under a nineteenth-century ethic was once admiringly hailed as "empire building," has impoverished many interior states and will eventually bring a tenfold curse upon the very powers that performed the exploitation. When the hinterland approaches desert status, the metropolis loses both its sources of nutriment and its economic power. The collapse of the metropolis follows upon the collapse of the rural community.

It used to be that predatory capitalistic elements chiefly threatened the welfare of the rural areas, and doubtless they still threaten, though in diminished degree. Now the great mass of farmers, small-town dwellers, small-business men, and the general unorganized public which Marxists call the "bourgeoisie" seem to be caught in a pincers movement, being hit from both right and left. They continually give of their substance and of their lifeblood, expressed in the youth they pour into the cities to renew hardened arteries and counteract the blight of race suicide.

In all of this a palpable social truth has been dismissed, but two factors now point to the inescapable need for readjustment. The first is that national economic and social stability cannot be maintained under the old method of doing things. And the second is that, whether we like it or not, the dawning atomic age will force vast changes upon us. It is a rude commentary upon American thinking that the economic and humane virtues of decentralization have had so little persuasiveness, whereas an instrument of destruction, the atomic bomb, made clear within a matter of hours the insecurity of cities throughout the land. The small town and the hamlet have suddenly become attractive to people who are prompted by fear. But they will acquire a status which is their due for quite other and better reasons.

Throughout all the confusions arising from fear and from the ambitious but often misguided blueprinting of social, economic, and political programs, we can still feel the steady pulse of the unique American adventure, whose strength provides the hope for a better era to come. In that era the individual, unfettered either by concentrations of economic power or by social directives, can realize himself and his country's best destiny. We shall achieve unity, if we will, through a wise decentralization which looks to the physical, spiritual, and economic welfare of the common man.

Made in United States
North Haven, CT
04 November 2026

21772150R00164